I0469508

Lethal Wind

By

Brian J. Spencer

This book is a work of fiction. Places, events, and situations in this story are purely fictional. Any resemblance to actual persons, living or dead, is coincidental.

© 2004 by Brian J. Spencer. All rights reserved.

No part of this book may be reproduced, stored in a retrieval system, or transmitted by any means, electronic, mechanical, photocopying, recording, or otherwise, without written permission from the author.

ISBN: 1-4140-3985-9 (e-book)
ISBN: 1-4140-3984-0 (Paperback)

This book is printed on acid free paper.

1stBooks - rev. 12/08/03

This work could not have been completed without the support and guidance of Eileen and John Spencer and my wife, Jennifer Sabo. I would also like to thank Ace, who with his long walks and unending patience, served as a fantastic sounding board.

Prologue:

July 1976

The screams were what haunted him the most. Those screams that crept down his spine like an ice-cold spider squirming down his shirt. He would wake in the middle of the night and hear the primal shrieks from across the room. Sometimes in the shower, he would suddenly look over his shoulder, certain that he had heard those cries from behind him.

At ten years old, there was nothing he could have done but hide. But after a lifetime of hearing his mother's screams, there were times when he wished he had stepped forward on that day and had allowed them to kill him with the rest of his family.

"Spank! Hey Spanky, you wanna go shoot some hoops?"

The voice startled him back to the present.

"Hey, Spank! Wassup?"

"No! I do not wish to play."

"Suit yourself." Jeff said as he pounded the ball against the lab floor on his way down the hall to the outside door.

The receding smack, smack of the ball was the only sound in the room tonight except for the usual sounds of the lab equipment. The hums and clicks of the various machinery working to keep samples cold or warm and working to clean the air so that no organisms could escape. How ironic that all this machinery was here to keep the organisms inside the confined spaces of the lab as though they had a singular purpose in life to escape and harm the people of the world. The only route of escape these organisms had was with the scientists that walked out of the lab each night unsuspected by the administration and security guards. It would be so easy.

Sa Reshdi carefully slid the thin gel onto a sheet of saran wrap and covered it. He took this into the dark room where he placed a large X-ray film on top of the gel. The radioactive particles in the gel would strike the film causing chemical changes in the silver compound, like light hitting the film in a camera. The radioactive particles corresponded with short fragments of DNA on the gel.

The clock ticked off the minutes as Sa waited for enough of the particles to bombard the gel. After ten minutes, he lifted the film off the gel and fed it through the X-ray film developer. He walked back outside the room to wait for the film to finish.

He lit a cigarette and walked to the end of the courtyard to watch the sun set over the Pacific Ocean. He had to make a decision tonight, but for these last few minutes, he wanted to watch the large red ball slide toward the ocean and its final resting spot. He had spent many evenings over the past two years at this very spot watching the wonder of nature's cycle, but beauty like this in a setting such as this were truly at odds with each other.

At over six feet tall, Sa stood taller than many of the other members of the Hoffman Institute; although, he constantly felt others were standing over him. Not all his time in Southern California had been bad, he had shaved his beard after arriving, and he found his dark skin was much more impervious to the effects of the sun than many of his co-workers.

The sun had slipped below the horizon of the water leaving the red and orange tinted sky as the only reminder of another day gone by. Sa turned and walked back to the lab picking up the developed film before heading back to his workspace. His bench, although much cleaner than his fellow scientists' benches, was still hard to qualify as neat. His lab notebook had a large brown stain on the cover; leaking into the first few pages from the time he spilled acid on his bench. The bottles lined up on the shelf above his bench were labeled with a multitude of colored labels only half of which were in English. He had learned long ago that it was better to label important chemicals and solutions in Hindi to prevent others from pilfering his hard work.

He was alone in the lab again as he had been for the past several nights. He was on a schedule to finish this project. Tonight was the last night. The film lay on the bench in front of him untouched since he had come back from his sunset. Fear led to procrastination. He was afraid of what he had created, and he was afraid of the decision he knew he had to make tonight. It was 8:00; he still had an hour before they would call and demand an answer.

Sa picked up the film and held it against the light. The DNA molecules had been labeled with a radioactive substrate that produced dark bands against the light tan film. Against the light, he could see a series of small dark bands forming a ladder from the top to bottom of the gel. There were four lanes of bands corresponding to the four molecules of which DNA was composed. By reading the gel from top to bottom he could determine the sequence of the DNA he had used. The sequencing reaction had taken him three days to set up followed by another two days to analyze the results with the gel and film.

Sa took the film over to a light box and clipped the film to it before turning on the light. Among all those dashes looking so much like Morse code, was the answer he was looking for. All he had to do was sit down and read this, and he would know if he had been successful; successful for his boss, the government and for his new friends. He just couldn't look at it yet. Sa got up from his chair and walked back across the lab, past the rows of study carols the scientists used for writing desks, past the water cooler and the coffee machine, and past the library with the all its journals stacked on shelves from floor to ceiling.

Sa sat down at Carolina's desk. He had been here many times, more in the past few weeks as he felt his time here coming to an end. Carolina had been one of the true bright spots in his time at the institute. The others in the lab could play their jokes on him and tell crude jokes, but at the end of the day, he had a person who cared for him and truly wanted to make his time in the United States fun.

Carolina arrived at the Institute from Brazil about six months before Sa. The two of them shared a background in virology and an interest in surfing. She had been a patient surfing teacher during the first few months, and he had tried to be a patient student. Once he really picked up the technique, the two of them would skip out of work on Friday and Saturday afternoons to surf the waves at the nearby beach. That was when they had started dating.

He had asked her to coffee one night after spending hours out in the waves. It had taken weeks for him to work up the nerve to ask her on a date and he could still remember how his heart pounded and his hands broke out in a sweat as he struggled to get up the courage to ask her for that one simple date. He had been intimidated by her beauty while sitting at the chair in the coffee house. Her long black hair, streaked a lighter brown from the hours in the salt water, hung past her shoulders. Her large brown eyes could peer into him to ask him the questions he didn't even know he wanted asked. She had made it easier for him to fit into this society. She understood him.

Sa stood and walked back to his lab. She understood everything except this. Carolina had come from a world of affluence and beauty. She wouldn't understand his background. She would never believe that he had watched as his mother was raped while his father was tied to a chair to watch. She would never have understood how he had watched as his village burned after all the adults had been slaughtered and then dumped into a large hole in the middle of town. The cries and whimpers from his neighbors as the dirt was being dumped on them still haunted him in his dreams. She would never have believed him if he had told her this. All this pain because the Americans didn't want to get their hands dirty in that part of the world. He had learned later in school in Delhi when he lived with his Aunt. How could he tell this woman all this? How could he tell her what he had seen?

No ten-year-old should have had to see that. The Americans could have prevented it. They could have stopped all that. Their president had known what was going on - ethnic cleansing. It wasn't until years later that Pakistan had become its own country. Now he had a chance to fight back and even the score. Only now he had to make a decision between his people and the one person he had ever loved. It was worth it wasn't it? They would call soon and ask him that very question.

Sa sat down at his desk and looked at the film. Carefully he traced his finger down the dashed lines as though he were reading Braille. He ticked off the sequence of A's, G's, T's and C's, reading the alphabet that made up DNA. It took him a half hour to read the three hundred letters before he reached the bottom of the film. As he stood up from his chair he felt the cold sweat of fear drip down his back and arm pits. He had been taught, cultivated and preened to do this very project from his start in the lab, and tonight he had just finished the last step. He should have been glad. Ready to celebrate. He wanted to call Carolina and share his joy, but not tonight.

They had asked him to create the ultimate weapon. Not one that would require a soldier to fire. Not one that would kill one, or a hundred or thousand people; one that would have the ability to wipe out a whole nation or race. Ironically, one that could fulfill the task of ethnic cleansing.

The phone rang on the wall beside his desk startling Sa back to the present. He looked up at the clock. It was nine o'clock already. Slowly he placed the film back on his bench as the phone rang again. His feet betrayed his emotion as he trudged for the phone and with a shaking hand, reached for the receiver. The phone rang again as Sa watched his hand as though from a distance. He watched as the hand lifted the receiver and pulled it close to his ear.

"Sa?"

Sa heard the voice. The now familiar, yet spine tingling voice of his contact. "Yes?"

"Do you have the virus?"

He had the ultimate weapon sitting on his bench. He thought about his parents, and his mother's screams as she was forced back at knifepoint. His father's shouts as he struggled against his ropes to try to free the woman he loved. He remembered the look in his mother's eyes when she saw him peeking around the corner and then her strange silence as the knife sliced across her throat. Sa's hands shook again and sweat burned his eyes.

He looked around the room he was standing in. The state-of-the-art equipment he had been given access to. The desks of the few friends he had made in the lab. Jason's desk with his stacks of journal articles and Heather's desk with her mirror and picture of her boyfriend taped to the wall behind. He looked at John's bench with its confusion of equipment and half-completed experiments. He looked at his desk and the picture of himself and Carolina standing at the beach proudly holding their surfboards.

He knew he was prepared to give this all up. A lesson had to be taught. They couldn't be allowed to do to others what had been done to him. Sa straightened up and wiped his sleeve against his forehead.

"Yes."

Chapter 1:

November 2002

His heart was pounding with fear, sweat was forming and running down his back and yet he knew he couldn't give up. The last fall had skinned his knee under his pants, but he wouldn't let his dad know that. He wouldn't give up until he could show him that he was old enough to handle a bicycle.

"Remember, keep your head up and pedal faster," his dad's voice seemed to be right next to him.

"Dad?" Jack looked down and saw that he was sitting on the seat of his motorcycle except that there was a sidecar attached to it now. His father was sitting in it.

"Dad? What are you saying?" Jack asked his father.

"I know you can do it if you keep your head up. You won't fall this time."

Jack looked around and saw that he was riding his bicycle again, but it was raining now.

"Dad! Dad! Where are you?"

It was getting dark out now and Jack knew he shouldn't be out after dark. He didn't want to make his Dad mad, but he didn't know where he was. This looked very unfamiliar. It was raining now and he couldn't seem to wipe the water from his face fast enough…

"Damn," Jack bolted straight up in bed squinting his eyes against the bright sunlight pouring in through the window across from his bed. It had been a while since he had dreamed of his dad, although he now knew what the rain was. Jack rolled his head to the right and stared right into the wet tongue of Ace.

"What's up Ace? Do you have to go outside?"

Ace was a large black dog that looked like a Labrador retriever crossed with a wolf. When he walked, his rear end swiveled independently of his front as though his spine was composed of a large universal joint. At 80 pounds, he could be quite intimidating and in fact his greeting was half way between a dog's growl and a wolf's howl; however, Ace was the gentlest dog to those who knew him.

Jack stepped into his slippers and grabbed his robe as he walked to the front door of his apartment to let Ace out. It was tough keeping a dog in an apartment in the city, but Jack had used his inheritance to buy the cabin outside of town. Besides, Ace didn't seem to mind as long as Jack was there.

"You know you could make my life a lot easier if you learned to use the toilet like the dog we saw on TV the other day."

The sound of a door opening behind Jack startled him. Samantha stepped out of her apartment next door to Jack's. She was wearing a simple black blouse and boots that disappeared under a long tan skirt. Every time he saw her, he felt his heart skip a beat. He was still trying to work up the nerve to ask her on a date, but the moment never seemed right. There was also the problem that he often saw another man stopping over at her place, particularly in the evening.

"Hi, Jack," Samantha glanced down the sidewalk with a puzzled look. "Who were you talking to just now?"

"I was, umm, I was talking to the dog." Jack stammered and quickly pulled Ace's leash toward the open door. "Good morning though," Jack called as he stepped back into the safety of his apartment.

"I've got to stop talking to you so much, Ace," Jack said as he filled the dog's food bowl. "I need to get out more. I need a date."

Jack knew the problem wasn't attracting women. He had had plenty of dates; however, most of them ended up as a first date only. At six feet and 165 pounds, okay 175, he wasn't your typical scientist nerd, but he also would never be mistaken for a football player. The problem he had dating was that he needed to find someone who would understand his long work hours and his occasional need to be alone. Right now, though, his research consumed him and didn't allow much time to meet people outside of science.

Chapter 2:

After a shower and shave, Jack looked at the clock —
8:30 already! He was going to be much later than he
wanted. Jack walked around back of the apartment
building to the parking area, past his beat up Jeep to
his motorcycle. Three years of hard work had been put
into the restoration of the 1959 Harley Davidson, and
every morning that Jack rode into work, he felt every
moment was worth it.

Jack inserted the key into the ignition on the left and
sat onto the well-worn spring cushioned seat. He
pulled the kick-start lever out of the right side of the
engine while he went through his mental start-up
checklist. With one final listen to the morning birds
and breeze, Jack hopped up out of his seat and dropped
with all his weight on the kick-start lever releasing the
beast within the large V-twin engine. Within seconds,
the roar of the engine quieted down to the familiar tha-
thump, tha-thump of the idling Harley Davidson. Jack
listened acutely to the engine for anything out of the
ordinary before easing out the throttle and pulling out
onto the street.

The ride into the Hoffman Institute was only a couple
of miles up the hill, which was hardly enough to enjoy
on the motorcycle, but Jack would take whatever he
could get. Some mornings he would purposely take
the long way into work to feel the wind and open road
a bit longer and delay the stale air and closed walls of

the lab, but not today. He was already later than he wanted to be.

The Hoffman Institute was a non-profit research institute that was dedicated to the study of infectious diseases. It was a small facility with only about 200 scientists; however, it was one of the most prestigious places to work on infectious diseases. Jack always felt lucky he had secured a position as a post-doctoral researcher here. He knew he could learn a lot working in this kind of environment and besides, John O'Leary, his boss, pretty much let him work on what ever crossed his mind as interesting. This morning though, he had a rare meeting with John to discuss his current project.

Jack walked into the laboratory and crossed in front of John's office. Glancing through the window of the office door he saw another man in there talking to Professor O'Leary. Although it was not unusual for the professor to take meetings early in the morning before he became busy with his numerous committee meetings, what sparked Jack's interest this morning was the visitor himself. The man appeared to be wearing a military uniform. The stranger was standing in front of the professor's desk, towering over the professor and having an animated conversation. Jack couldn't determine individual words, but the stranger was talking loud enough so that his voice could be heard through the closed door. Jack got the impression this was a man used to giving orders to large groups, and those orders were to be obeyed without question.

Jack walked to his own desk and dropped his bag down.

"Hi Jason. Is there fresh coffee this morning?"

"Fresh?" Jason looked up from the pile of papers he was reading at his desk. "I'm sure it isn't fresh and I'm not even sure it's coffee, but there is some hot liquid with caffeine in the pot so give it a shot."

"Hot caffeine, that's all I ask," Jack said as he grabbed his mug and walked over to the coffee maker.

Amy was standing at the coffee maker in her trademark jeans and 80's British punk rock band t-shirt. This Monday was apparently a Public Image Limited day. "Good morning Jack," she said in her British accent. Amy had come to Dr. O'Leary's laboratory after receiving her Ph.D. at Oxford University.

"Don't tell me you took my coffee," Jack winked at her.

"Your coffee? Hah! I've been here an hour already. I think I definitely deserve this. And just like you, I'll leave an empty pot for you to suffer over." She poked him in the stomach as she exaggerated a strut while walking by.

Amy and Jack frequently teased each other about coming in earlier than the other. More often than not it was Jack that made it in to work first. He felt pretty sure she woke up earlier than he did, but she obviously spent quite a bit of time every morning making sure she looked good before leaving her apartment. Amy had the kind of hair most women dream of having. Little did they know that she spent at least an hour every morning curling it and drying it so it looked just right. These are the sorts of things you learn about a person when it is 2:00 AM and you are the only two people still in the lab.

Jack poured new grounds into the coffee maker and loaded the water. He was staring off into space waiting for the pot to fill when he heard his boss, John, calling for him. Jack looked up and saw that John was standing alone in his office door. The army suit must have left while Jack was making coffee.

"Jack, I need to talk to you for a bit. Do you have a second?" John asked.

Jack poured himself a cup of coffee and grabbed his note pad before heading into John's office. It's fascinating that some of the top medical research can come from some of the most unorganized investigators. John's office was small for someone who had published as much as he had and had worked his way up from a post-doctoral researcher at this same institute to head of his department. He felt comfortable where he was and routinely turned down offers of a bigger office. But a small office meant very little space for journals, textbooks and visitors. John's desk took up nearly half the room and was situated so that it faced the door. The entire left hand side of the desk was piled approximately two feet high with the past three years worth of journals on subjects ranging from biochemistry to gene therapy, and he had read them all. Even more amazing was that he could pick a specific journal out of the stack and practically turn to the page of the article he was making a point about.

Jack grabbed a seat in front of the desk. He wanted to ask John who he was talking to before, but he knew John was the kind of person that would easily brush off such questions. Although Jack felt comfortable around the renowned Dr. O' Leary, he was a bit concerned about a scheduled meeting with his boss.

"Jack, I just wanted to touch base with you since I was out of town last week. I want a progress report on your research."

"Well, good morning to you too, John." Jack knew that the use of his first name by a post-doc really grated on the professor, and he suspected had anyone else tried it, they would have heard from him on the topic. Jack often treated superiors as equals rather than authority figures, sometimes to his detriment.

"Jack," the professor said with not a little impatience in his voice, "let's get to the point here."

"Well, I do have some good news about my anti-viral project." It was always a good idea to save a few good results up your sleeve for times like this when your boss demanded results. Science was fickle, not like the movies or TV showed. An investigator could spend weeks or months working on a project with failure after failure. Those were the times when one most dreaded the "progress report" talks. And then one day, bang! Something would break. This was usually short lived though and was often followed by many more failures. Unfortunately it was only those big breakthroughs that made the morning or evening news, so most people never realized how much time and energy could be spent on a project.

"As you may remember, my direct goal in the project has been to block the cold virus, Adenovirus. I have been testing various molecular compounds that would block the replication of the virus by specifically targeting virus proteins. Last week I found one compound that was able to bind to that protein and block its action and thus the replication of the virus." Jack paused.

"Well, that is good news. I know sometimes these projects can seem hopeless, but I have a lot of faith in you, Jack. So what's next then?"

"Well, I have data showing the compound can block the virus replication in a test tube. Now I need to find a way to deliver the anti-viral compound to the nasal cells. Since the Adenovirus infects primarily the nasal cells and upper respiratory tract when it causes the cold, I need to deliver the drug to those sites. An injection under the skin or into the blood won't stop the virus fast enough."

Brian J. Spencer

Jack was the expert in the lab in viruses, so he often felt the need to explain the virology behind his project to the professor. Although he was certain the professor's extensive knowledge did include some virology, he suspected John let him explain things simply so that they both knew they were on the same page when the discussion turned to more specific ideas.

"Have you considered fusing the compound to a transporter protein?"

"I know where you're going with this. I worked on those transporter proteins a bit during my Ph.D. thesis at the University of Washington. Attaching those small specific proteins to this anti-viral compound would allow the drug to enter the cells of the airway if we delivered it as an inhalant spray. I'm going to make those new compounds next and then test them on the mice I have. I really think this is the break we've been looking for in this project."

It was sometimes hard to keep a project going long term in the lab, especially when one was in Jack's position. As a post-doctoral researcher in the lab, he only had a three to five year span to work, and then he would have to find a more permanent position somewhere else. In that time he needed to get papers published on his work. But he wasn't the only one who had to be excited about a project like this. His boss, the principal investigator of the lab, also had to see the promise in his work and support him financially and scientifically.

"I have seen your past work, Jack, and I have complete faith in you. And I am sure that your funding will be renewed next year when you publish these results. This really is good news. I guess I don't really need to check up on you regularly. On your way out can you send in Amy please?"

Jack got up to leave and then paused. "John, who was the army suit in here talking to you before?"

John's face turned a bit darker and the crease lines around his eyes disappeared. "Let's focus our discussions on what is important to your work here. I will take care of the administrative portion of this lab."

And with that, Jack knew his audience with John was done for the day. He was a bit sorry he had soured the discussion with the question; even more sorry that he had to send Amy in now. He should have just shut up and let Amy have her talk when he was still in a good mood.

Chapter 3:

Jack looked at his watch. Wow! It was almost 11:30 already. He had lost the whole morning. He started toward his bench intending to get some work started before lunch.

"Amy, the boss wants a word with you. And hey, do you wanna go get some lunch after you're done."

Amy gave him the thumbs up sign before disappearing into Dr O'Leary's office. She was learning these little Americanisms from Jack.

For the next hour Jack worked at his bench, but his mind kept coming back to his talk with John. The professor had seemed a bit overeager at the news of the anti-viral research. Although it was not unusual for him to become very excited about a project, he seemed to be spending extra attention on Jack's work lately.

The professor had made his name in the field through his work on toxins, in particular the ricin toxin. He had been the first to isolate the toxin gene from the castor bean plant and had cloned this gene for further analysis. He went so far as to identify the mechanism by which ricin acts on cells and kills them. It was this research that guaranteed him a position at the Hoffman Institute.

Most of the lab now worked on cellular proteins that were involved in chromosome replication. This branch of his career evolved naturally from his work with the toxin because the proteins that the toxin affected were also involved in this same process.

Now the only person working on ricin was Amy. Jack's research seemed the most far-removed from everyone else in the lab. He was the only person working on an anti-viral drug. When he had first joined the lab, John had persuaded him to take up this line of research as a new avenue for the lab to follow in the future. John had explained that he had seen to many of his colleagues passed by in science for their failure to apply their knowledge to new fields and broadening their appeal to funding agencies.

"Hey Jack, you look deep in thought," Amy was standing right beside his bench. "Do you want to go to lunch now?"

"Oh, yeah. How was your talk with Dr. O'Leary?" Jack asked.

"It went pretty well, but I'm really hungry so let's go get food and then we can talk outside."

As they walked by Dr. O'Leary's office, Jack saw that the office door was closed and John was on the phone having another animated discussion.

"Didn't he want to talk to anyone else today?" Jack asked Amy.

"No, he said he was going to work alone today and that he would probably be gone tomorrow through the end of the week."

"Strange, I had gotten the impression he wanted to check up on everyone's research but then he only talked to you and me."

Jack and Amy bought lunch at the small cafeteria that was associated with the Institute and then headed to one of the tables outside. It was another gorgeous day out with sun high overhead. Lunch was one of the few times most of the scientists in the facility got to enjoy the life of living in Southern California so the tables were usually quite full. Jack found a small bench off to the side and motioned for Amy to join him.

"So did he like what you had to tell him?" Jack asked Amy now that no one was around to overhear.

"Yes, quite so. I told him that I had finally decided that developing a vaccine against the ricin toxin was not going to work and that I had moved my research into the area of trying to block the toxin directly."

"Why won't a vaccine work? We have vaccines for the toxin that causes tetanus."

"Tetanus toxin is produced from the bacteria following an infection in the skin. Like when you step on a rusty nail. The toxin is produced slowly at first with more being produced as the bacteria grows. This lag in toxin production allows the body a chance to mobilize an antibody response against the toxin. So the vaccine merely primes the immune system to build up these antibodies in advance.

"If we are going to find a treatment for inhaled ricin, we need to block the action of the toxin immediately. The thought is that if this toxin were to be used as a biological warfare agent, we need to mobilize a response to the toxin very quickly.

"I tried to develop a vaccine to test in mice but even those animals that were immunized against the toxin and had antibodies to it, died when exposed to the inhaled toxin. So now I am directing my efforts at blocking the action of the toxin by developing a protein that will bind directly to the toxin."

"And Dr. O'Leary liked this approach?" Jack asked.

"Yes, indeed. He even asked me to talk to you about it. He said you are trying a similar approach with your anti-viral research."

"Yeah, that's true. I can show you my work this afternoon and hopefully it will be helpful. So, do you find it odd that he only wanted to talk to the two of us and not the rest of the lab?"

Amy finished a bite of her sandwich. "I was thinking that maybe it has something to do with the man we saw the professor talking to this morning. Remember, we are both funded from a grant from DARPA."

"DARPA?" Jack asked her.

"Yeah, the Department of Advanced Research Projects Agency. It is a research funding division of the Department of Defense. They fund projects that may have some future meaning to the country's security. They funded the development of the internet."

"Really the internet?"

"Yes, the Defense Department wanted a method for quick communication with its various bases across the United States, and so out of that came the internet. I helped the professor write a grant for DARPA right before you came here. The project is involved in developing novel technologies to combat bioterrorism."

"I never looked at the grant I was funded from. But what does working with the Adenovirus have to do with bioterrorism. I mean its not as if the bad guys can really hurt us by giving us all the cold."

"The adenovirus has much more potential than just as an agent of the common cold. You know it is the most developed virus for gene therapy. It doesn't take too much effort to see that if we can use the virus to deliver genes to treat genetic diseases, one could also use the virus to deliver much more deadly genes."

This was a lot to digest. Jack and Amy ate the rest of their meal in silence. Afterward, they walked back to the lab where Jack got another cup of coffee. The rest of the afternoon went smoothly for Jack. He was thinking about his conversation with Amy. He hadn't even realized that he was working on a potential bio-weapon. The common cold was not at all serious and wasn't even a glamorous project, but it did have a serious economic impact. Just like chicken pox, the common cold forced people to stay at home for a few days missing work and reducing productivity. Parents stayed home to care for their children when they invariably came down with the annual cold. In fact in terms of lost time and productivity, the cold was far more devastating to a developed nation. On top of this was the fact that the cold virus was constantly changing so that even if you had recovered from one cold, you could easily "catch a cold" the next year.

At 6:00 Jack finally realized he wasn't going to be very productive anymore. He walked out to his motorcycle and started it up listening to the sound of the engine tha-thump, tha-thump. As he was putting his helmet on he saw a very nice silver Nissan 300ZX drive past him and then turn around and park across the street. He noticed nice cars often. In fact he could pretty much name the driver of all the nicer cars in the parking lot.

Since it was still early, Jack decided to ride home the long way. He rode his bike up the coast for a while watching the sun set to his left. At one particular stretch of the road, you could look out over the ocean while standing on a cliff. Jack stopped there and watched the surfers lying on their surf boards just beyond the breaking waves. Every so often, one would start paddling and would catch a wave riding it in for a bit. The waves seemed small today, but what did Jack know, he had never even tried surfing. For some people, salt water was just in their blood. Jack had grown up in Wyoming where the only salt water was the liquid you gargled to get rid of a sore throat. No, Jack had fresh water in his veins. One day he hoped to get back to an area of the country where fresh water was so abundant he could do some bass fishing again.

Jack sat down with his back to the motorcycle to watch the red and orange sky turn to purple and finally black. Although long stretches in the lab often wore on him with the continuous failures, he still felt he enjoyed his work. As long as he enjoyed going into work each day and it didn't start to feel like 'work' he felt he was in the right field. He felt lucky to have chosen this career to work in since he enjoyed it so much. With the setting sun came a cool breeze that blew through his leather jacket letting Jack know that night was coming. Ace would be waiting for him back at his apartment.

Chapter 4:

Jack rode the rest of the way home in the dark. As he was pulling up to his apartment, Samantha was just walking out of her door. He took his time removing his helmet and shutting down his bike as he looked at Samantha. Even in casual jeans and a t-shirt she looked gorgeous. She must have been at least 5'10" with short, red hair that was so tightly curled it didn't seem to move when she turned her head.

"Hi Jack, I was hoping to catch you tonight."

"Really? I mean…great, well here I am." Jack could feel his heart rate step up a gear at the sound of Samantha's soft voice. Her large brown eyes looked as though they took in more than just the view in front of them.

"I have a motorcycle question actually. Can you spare a few minutes?"

Wow, gorgeous and interested in motorcycles! Could this be a dream? "Sure, let me just let Ace out quick and give him his dinner. How about 5 minutes?"

"Okay, just knock when you come over." Samantha said and then turned and walked back into her apartment.

Jack entered his apartment and greeted a very happy Ace. "Hey buddy, are you hungry? You're never

27

going to believe what just happened." Jack continued talking as he got out the dog food.

"Samantha invited me over to her place. Okay so it was just to answer some questions about motorcycles. But still, it's a start isn't it? I haven't seen that other guy over there lately. Maybe he is out of the picture."

Jack went into the bathroom and brushed his hair quickly. One of the problems with riding a motorcycle everywhere was that your hair was often matted down from the helmet. It was worse than riding in a convertible.

"Okay, you take care of yourself for a bit. I am going next door," Jack called to the dog as he left the apartment.

Samantha let him in after the first knock. Although the floor layout should have been the same for both apartments, Samantha obviously was much more adapt at decorating so that it looked as though she had quite a bit more room. Jack walked directly into the living room that was connected with the kitchen by a breakfast bar. Looking around the room, Jack saw a large couch with a coffee table in front of it. On the coffee table was a whole home office of computer equipment. He saw a large monitor and laser printer and then he noticed the computer console next to the table. He also noticed a lack of a TV in the room; almost a staple in every American living room. On the wall across from the sofa hung a large picture celebrating the bands that had played at Woodstock.

Not the new and trendy Woodstock of the '90's, but the original Woodstock with Jefferson Airplane, Bob Dylan and Jimi Hendrix. This was a woman he could relate to.

While the lack of TV and the presence of a home office in the living room had caught Jack's attention, it was the object in the hallway that most fascinated him. A large sheet had been draped over what was obviously a motorcycle, although Jack could not see what kind it was.

"Sooo," Jack started, "Have you got any motor oil, er I mean beer?"

Samantha followed Jack's gaze to the hallway. "Do you want that shaken or stirred?" Samantha asked.

"Uhh…"

"I was joking, I can get you a beer, and in the mean time, why don't you take a look at what I have there. I can see you aren't going to hear a thing I have to say until you get a look."

Jack walked past the kitchen entrance to the hallway and pulled the sheet off the motorcycle. A low moan escaped his throat, and then he whistled to show his appreciation at the beauty that was standing before him. The sleek, glossy black gas tank sat perched high on the bike and was followed by a beautiful black vinyl, straight seat. The exhaust pipes gently curved out each side of the engine to exit near the rear wheel.

Each wheel was put together with hundreds of shiny chrome spokes. Every curve and angle of this motorcycle had been hand-pounded and put together in the factory when new. The care and precision of those craftsmen still showed today.

"Wow, a Vincent Black Lightning!" Jack leaned in, "And with only 19,000 miles. This bike is in excellent shape. I'm very jealous." Jack looked up at Samantha who was holding two pints of beer, "It's kind of like that song by Richard Thompson 'There ain't nothin' in this world beats a Vincent and a red headed girl'." Jack blushed as he realized he had just said that line about Samantha.

"Here's your beer. It's not my motorcycle. My brother brought it by and asked me to store it for him for a while. I'm kinda worried about him so I asked you over to tell me about the motorcycle. I don't know too much about them but this one looked really nice. Is it expensive?"

"Well, I'm no expert on these bikes, but I do know they were made up to the mid 50's. This one looks original and in very good condition. With low mileage like that, I would have to say it's probably worth thirty to forty grand easy."

Samantha sighed and walked over to the couch and sat down. She threw her head back and covered her eyes. Jack could here a soft whimper from Samantha and could tell she was crying. He sat down next to her on the couch.

"I'm sorry Jack, it's just what I was afraid of. I don't want to put my family troubles on you."

Jack looked at the woman sitting beside him. From this distance he could tell she wore almost no makeup with only a light red lipstick. Even crying, her eyes seemed to invite him to her. "Well, I really didn't have anything planned tonight. Why don't we go out get some dinner and if you feel like talking about the bike, you can. Or we can just talk about other things. What do you say?"

Samantha reached for a tissue a dabbed at her eyes; "Sure, I could get out of here for a while. I just keep looking at that motorcycle. What did you have in mind?"

"Well," Jack said, "there is a nice Thai place down the road I have tried before. We could even walk there. It's really nice out tonight."

Samantha grabbed a leather jacket to throw over her t-shirt and locked her door.

Chapter 5:

"We've lived next to each other for over a year a now and I don't even know anything about you except that you ride a motorcycle and talk to your dog, Ace."

"Well, what do you want to know?"

"For starters, where do you work?"

"Well, I am a post-doc at the Hoffman Institute, and I work on developing anti-viral therapies for the Adenovirus…"

"Wow, slow down a bit. What's a post-doc? And what kind of virus was that? Is it dangerous?"

"I guess I have been spending too much time around scientists. Sorry. My job title is Post-doctoral Researcher. I earned my Ph.D. in cellular biology from the University of Washington two years ago. Now I work long hours for very little money in order to get more experience in the field. It's kind of like an MD doing a residency after medical school. After this job I could go on to get a permanent position somewhere. Right now I still work for a professor who acts as the principal investigator on all my projects.

"Now then, the Adenovirus is one of the viruses that accounts for the common cold. Most people know that viruses cause the cold, but what they don't understand is that several viruses actually cause the same

symptoms.   So the Adenovirus is just one of them.
And actually there are many viruses that all fall into
the category of Adenovirus. The name simply implies
a category for viruses that all have similar genes and
structure."

"So why are you trying to make a drug for the common
cold?   Aren't there many more deadly viruses out
there?   Like what about HIV or Ebola?" Samantha
asked.

"Ah, here is the restaurant. I'll answer that question
during our dinner."

Tim's Thai Place was a very small restaurant with
probably only 15 tables. Tonight only five tables were
occupied and those by couples that all looked like they
were in their 50's. A short, 50ish looking Asian man
walked up to the front to greet Jack and Samantha.

"Hi Jack, I was hoping you would stop by tonight. I
wanted you to take a look at my car," the man said. He
glanced over Jack's shoulder and back at Jack. "Oh,
I'm sorry, you're here with a date."

Jack blushed.   "No, this is my neighbor Samantha.
Samantha this is the owner Tim."

Tim was obviously embarrassed as he wiped his hands
on the towel he had tucked into his back pocket before
reaching out his hand to Samantha.

"Nice to meet you, Tim. I've lived in the neighborhood for three years now, and I don't think I have ever seen this place before."

"I've been here for 22 years. Seen a lot of changes in the neighborhood, but I still mostly get my regular customers. Jack here is probably one of my youngest. But we need more pretty women like you to class up this joint. Jack, the car can wait. You can sit anywhere you want."

Jack led Samantha to a table along the wall and pulled out a chair for his guest. A small light hung low over the table providing the table with a dim glow.

"Tim has an old Volkswagen Kharman Ghia," Jack explained to Samantha. "The car is always having trouble, and so I help him keep it running. He usually gives me dinner in exchange. Guess I'll have to pay tonight though."

"Nonsense," Tim winked at Samantha "I think I owe you many, many dinners. Besides, I should give you a free dinner just for bringing this lovely lady here."

After Tim left, Samantha leaned into Jack and whispered to him, "Tim seems an odd name for a Thai man. And he doesn't even have an accent."

"Tim's a third generation American. His grandparents came over here during the Thailand depression in the 20's. He is all American. But he still can't fix a car. So where was I before?"

"You mentioned something about the cold virus Andeno-something?"

"Adenovirus. I know it isn't one of the sexy viruses. It won't make the cover of Time magazine and I won't have an interview in the New York Times for working on it, but this virus does contribute significantly to lost work time. Employees with the cold typically spend 2-3 days at home recuperating and employees with children will often have to spend even more time at home. That all amounts to a lot of lost productivity. A vaccine for chicken pox was recently developed for this same reason.

"There are other reasons for working on this virus, too. It's one of the key viruses used in gene therapy for delivery genes, and because it is a relatively simple virus to work with, it has the potential to be a bio-weapon."

"What! How could someone use the cold virus as a bio-weapon?" Samantha asked.

"Actually the same way it could be used for gene therapy. A virus has the unique ability to enter cells in the human body and deliver genes to that cell. In the normal life cycle of a virus, those genes are for the use of the virus and they involve replication of the virus, or copying itself. They are also used to block an immune response against the virus so it can stay in the body longer and potentially spread to other cells or other people.

"In gene therapy, the strategy is to use the ability of the virus to deliver genes to cells to deliver a correct gene to cells that have a defective gene."

Samantha interrupted, "But won't those people just get sick from the virus?"

"Actually, I skipped a step. Before we put the good gene into the virus, we remove all the virus genes. So the virus becomes a vehicle to deliver the good gene without the ability to replicate or cause disease. We are just using one of nature's tricks for our cures."

"So then tell me again, how could the cold virus be used as a weapon?"

"The same way we use it for gene therapy. The virus already has the ability to infect humans and spreads easily among humans. If someone wanted to keep those abilities of the virus, they would leave those genes in. Now we know the cold virus won't kill you, so to make the virus a weapon, all you would have to do is insert a 'killer' gene. Something that would kill you like the botulism toxin."

"Wouldn't it just be easier for a terrorist to release the HIV virus or Ebola since those already do kill people?"

"Those viruses are actually poor weapons. The HIV kills slowly, and we have some effective treatments for it already with many more in development. Ebola kills too quickly so there isn't a chance for it to spread.

Besides, Ebola mutates so fast in the humans that usually the virus mutates to an ineffective killer after only a few weeks. That's why those infections in Africa don't last very long and don't spread very far. The advantage of the cold virus for a weapon is that it spreads fast and it has been well studied so scientists know what genes are necessary for each step of its life cycle."

Samantha ate her food in silence for a while apparently lost in thought over the conversation she had just had. Jack watched as she pushed peanuts around her plate trying to balance them on the tines of her fork. She really seemed to like the peanuts but she kept pushing the water chestnuts to the edge of her plate.

Jack wanted to talk about the motorcycle in her living room. The topic of the bike had clearly upset her. It must have something to do with her brother. Samantha finished the last peanut and left a plate clean of food except for the water chestnuts.

"So, I guess this Adenovirus really is a lot more important than the common cold. How did you get into working on viruses?"

Jack really wanted to learn more about Samantha, but he didn't want to push her. Besides he was having fun just sitting here talking to her.

Chapter 6:

"Do you want to get a bottle of wine and just talk here for a while?"

Samantha looked at her watch quickly. "Well, okay, but I really should be home by 11:00. I have to be at work early tomorrow."

Jack called Tim over and asked for a bottle of Zinfandel. He settled into his seat to tell Samantha his life story.

"My Dad was a writer while I was growing up. He wrote mostly historical fiction about life in Europe and America around the 16th and 17th centuries. He had gotten his degree in history but then couldn't find a job in his field so he just started writing. He was pretty good and had 12 novels published. I remember I was about 10 years old when I first realized he was a writer. I guess before that I really didn't know. He always went to the office each day just like the other dads. Only his office was a small room in the basement.

"For my 10th birthday, I had a party with all my friends and my dad gave me my first bicycle. He also gave me a leather bound copy of his latest book which he had dedicated to me. It was way beyond my reading level, but I finally managed to read that book all the way through. I remember thinking it was bigger than the bible and I would never get through it. I swear I had to look up every other word in the dictionary we had in the living room. Anyway, this book was about Louis Pasteur and his search for the agents that caused food to spoil. It was his search for bacteria. After Pasteur, people knew to sterilize food and cooking surfaces. His studies greatly reduced the spread of microorganisms at a time when smallpox and the plague were rampant. The story was told from the point of view of an assistant of Pasteur who respected him and followed him everywhere. But this was the first time I believed that someone studying biology could change the world. I knew from school that people could change the world through war. Hitler certainly had. And people could change the world through peace. Woodrow Wilson had. And people could even change the world through inventions as Thomas Edison had. But I never dreamed that you could change the world through biology. But Pasteur had done just that.

"It took me almost a month to get through that book the first time. But I remember," Jack paused and stared off to the far wall lost in his thoughts, "I remember, that is when I knew I wanted to be a scientist. At the age of 10, I knew I wanted to study biology and change the world. So," Jack looked back at Samantha, "that's what I studied all through college and went right to graduate school to study. And here I am, I haven't changed the world yet, but I am a scientist. The thing is, I think Dad just gave me the book because he was proud he had dedicated it to me. I read it and thought he gave it to me because he knew I wanted to be a scientist before I'd even realized it myself. I spent the last 20 years trying to be the scientist that I thought he wanted me to be."

Jack closed his eyes and remembered those days in Montana. Since Jack's mom had died during childbirth, and he was an only child, it was just the two of them his whole life. Looking back now he realized he was so proud of his dad, the accomplished writer. Being able to write from home had left him lots of time to take care of Jack. He coached Jack in little league, and even though he was arguably the worst player on the team, he never heard about it from his dad. His dad was there to teach him how to ride a motorcycle around the dirt trails when he was young, even though Jack knew he could never race on the circuits like his dad had when he had been younger.

"I think I spent my whole life trying to live up to an ideal that I thought my dad had of me. I worked so hard to be good at everything he was good at. It wasn't until after he had died that I realized that it was all in my head. He never set those standards for me. I only imagined that he did. Looking back on it now, I know that he would have supported me if I had told him I wanted to skip college and just be a mechanic."

Samantha finally spoke up, "I think your dad would be proud of you now. Maybe he really did see something in you at an early age that made him think that a book on Pasteur would be of interest to you. I mean after all, it was that book that he dedicated to you. Not another one."

Jack looked around and realized he and Samantha were the only ones left in the restaurant. Tim had moved to the back room to allow them to talk. He had put the radio on and Jack could hear Mr. Sandman playing softly. The wine must have loosened him up because he never talked about his dad that much. It was just easy to talk to Samantha.

Samantha interrupted his thoughts. "I suppose you want to hear about the motorcycle now."

"Look, we're having a great time out tonight. I don't want you to talk about it if you don't want to."

"I think I do. It's funny, I have a lot of good friends here in the city, but you're the one I want to talk to about this.

"It belongs to my younger brother Brandon. You may have seen him. He often comes by my place for meals, because he can't really cook. He's 25 years old but I still feel I need to take care of him. He moved out here two years ago and hasn't really had a job since he arrived. Oh, he has money, he doesn't leach off me. But it's where he gets that money that has me worried. I ask him about his job, but he always seems to be surfing or off-roading with some vague friends that I've never met. And then two days ago he dropped off that motorcycle and said he had just bought it, but he wanted me to keep an eye on it for a while. He told me he didn't have anywhere to park it for now."

"Where do you think he gets the money from? Is he in trouble?"

"I was hoping you would tell me that the motorcycle was a fake or that it was a cheap new motorcycle. I guess I already knew it was neither. I know he couldn't have come up with the 30 thousand to pay for it. So, yeah, I do think he is in trouble. I know he's often not at home when I call at night after work. He has to return my call during the day. And then last month he was out of town for a whole week. When he returned he wouldn't tell me where he had gone and even yelled at me for trying to control his life. He never yells at..." Samantha started to cry again and got up to get a napkin from the bar to wipe her eyes.

Jack got up too and put an arm around her shoulders to comfort her as she turned and buried her head in his chest.

"I'm sorry, I finally get a chance to meet you and I am such a mess. I should have invited you over weeks ago when my life was simpler. Then you could have gotten to know me when I was normal.

"It's just that I really do think he is in trouble, and you just confirmed it by telling me how much that motorcycle is worth. I know he can't afford that. I just don't know what to do. God, Brandon, why do I have to take care of you!" Samantha yelled at the empty room.

"Sometimes he's being dropped off at my place by some guys in a really big Mercedes. One time one of them got out to get something out of the trunk to give to Brandon. This guy had dark skin and a dark beard and was wearing a really expensive-looking suit. I was worried then that Brandon had gotten mixed up with something bigger than he knew. But he won't tell me anything."

"I know you probably don't want to spy on your brother, but maybe you could follow him around for a day and just see what he does."

"I've thought of that, but he knows what kind of car I drive. I have a red Austin Healy. It's not exactly the kind of car a person can disappear in."

Jack could see he had just volunteered himself for a job he really did not want. What did he know about following someone around? He had never been what most people would call a type A personality. Danger was most definitely not his middle name. But there was something about Samantha that he really liked. He looked into those warm, brown eyes; eyes he knew he could lose himself in.

"I guess I could check up on him for a day. I have the kind of job where I could take off for most of the day and nobody would mind."

"Really, Jack, I don't want to put you in this situation. It's just that I am really worried about him and I don't want to go to the police about this. I can give you his address and phone number. If there is anything else you can think of that I could do to help…"

"Thanks, I can't promise anything. I've never done anything like this before. But I'll give it a try."

Jack and Samantha walked the rest of the way home in silence. Jack knew he was a sucker for a pretty girl. There was probably nothing she could have asked him to do that he would have denied her.

"I want to thank you for tonight Jack. I really needed to get out of that apartment and away from that sheet in the hallway. I covered it up so that I wouldn't have to look at it anymore. I just know he's in something bad."

Samantha leaned in close and gave Jack a kiss on the corner of his mouth and then walked into her apartment. Jack sat on his front step for a minute relishing what was perhaps the softest and most tender kiss he had ever felt.

He couldn't believe what he had gotten himself into. And he realized he still didn't know that much about Samantha. Well, that was just a good reason to ask her out on another date. Maybe he could take her to a nice dinner on Friday. Jack stepped into his apartment and greeted Ace who seemed very happy to see Jack back for the night.

Chapter 7:

The next day at work Jack spent his morning working at his desk catching up on paper work. He tried to throw himself into his work. The past few weeks in the lab had not been kind to him. Failure had been more the norm lately, which just served to spur him on harder. He felt he was on the verge of a breakthrough, but then again that was always the case in science.

Research was as addictive as gambling. You could work weeks or months without a single victory, a single success in the research. And then, one thing works. You could reach such a high from that one success. All the failures and head banging of the past are forgotten and your presence in the field is justified. There's really nothing like that high of having a big experiment work and justifying all the time you spent on it. It makes you work through all the failures and losses that will ensue in the coming months to just get back to that feeling. Yes, Jack thought, science is for the addictive personality.

Jack's work with the anti-viral drug for Adenovirus had been mired in that pit of failure for the past few months. Every time he thought he had a good idea or new avenue of research, it failed. He was so sure that this idea of fusing the anti-viral to the transporter protein would work. He would be able to justify his funding and get back to that high.

Right now though he couldn't concentrate on his work. All he could think about was his dinner with Samantha. Before last night the only things he knew about her were that she worked at the Union Times newspaper and that she liked to listen to Bob Marley CDs played very loud on her stereo. He had really enjoyed spending time with her last night, but now he was feeling like he had gotten in over his head. How could he have agreed to follow someone? Maybe it would work out to be nothing. Brandon probably just had a good paying job he was embarrassed to tell her about. But what kind of job would he not want to tell his sister about? Was he really in trouble?

An arm reached around his shoulder and produced a full steaming cup of coffee for him. He turned around to see Amy smiling at him.

"You looked like you were going to fall asleep at your desk. I know the boss is gone, but you still can't sleep here." Amy teased him.

"Sorry, I was just thinking of something. I did want to ask you if you could take care of an experiment for me tomorrow. I have to run some errands, and I may be out all day."

"Certainly Jack. Dr. O'Leary won't be back until Friday so don't worry."

"Where did he go now?" Jack asked.

"He said he had to fly to Washington to meet with the DARPA group."

"You know. I still don't understand this DARPA group and our funding. When I came here, I didn't even look to see where my funding was coming from."

Amy pulled up a chair. "You Americans are all alike. You don't care who is behind the money as long as you get paid."

Jack hated these generalizations, but Amy was always trying to compare all Americans to the English. She seemed to delight in the history between the two countries.

"DARPA stands for Department of Advanced Research Projects Agency. Their primary mission is to fund research that may have future implications in the field. Often times they will look over a whole field of research and compare that to today's problems and tomorrow's perceived problems, and then they will identify areas that they feel need more focus. The agency then will put out a call for projects that fit within that goal. This isn't just limited to bio-weapons threats. As I said, they funded the development of the internet. They also have funded projects in aerospace as well as mathematics and physics. I can't believe I know more about this than you do. This is your country, not mine."

"And so my research is supposed to identify a cure for the Adenovirus in the event that it is used as a bio-weapon?"

"Well, there are the obvious other benefits of your research. DARPA doesn't get everything you work on all to themselves. Just like any other funding agency, they just give you the money. You still publish all your results so what you accomplish is public domain. What you are working on is a treatment for a potential bio-weapon, but it is also a treatment for a serious health hazard that affects people on a regular basis. DARPA isn't going to take that away from you."

"But if I publish what DARPA funds, then those people that are creating these bio-weapons will know our strategy for stopping them. They could easily bypass our defenses based on the research that I publish. Remember these article are out there on the internet for anyone to get at. What's to stop these people from using my research against us, or worse, what's to stop the government from using my research to develop a better weapon?" Jack asked.

Amy sat there for a minute and stared at her shoes. "Well, I guess we just have to trust that we know best. We just have to work each day believing that what we are working on is really for the good. Besides, your president Nixon signed a treaty with the Soviets in 1972 to ban the development of bio-weapons. I know that doesn't stop other countries and we all know it didn't stop the Soviet Union from developing Anthrax through the 1970's, but we have to believe that the

American government wouldn't fund us to help their bio-weapons program." Amy looked Jack right in the eyes, "I guess we just have to decide for ourselves; are going to go through life believing people are good, or are we going to go through life never trusting anyone?"

Jack finished his coffee and stood up, "I guess you're right Amy, wherever the money comes from and for whatever reason I get it, I am still working to treat the common cold that affects all people. Thanks, I guess I have just been getting down on myself lately." Swinging his leather jacket over his shoulder Jack said, "I'm going to go out for a ride before it gets dark. Thanks for taking care of my work tomorrow."

"Jack," Amy called after him as he was walking out the door, "If you want to talk later, I'll be here for a while."

Amy was a good friend to have in the lab. When Jack had arrived a year ago, he had immediately latched onto the one person in the lab that he thought could really help him. She had shown him around the institute and had helped him find a place to live. The first few months here had been lonely since he had left his girlfriend and most of his friends in Seattle. Amy had been the one person here he had enjoyed going to dinners or movies with. They worked similar hours and had been each other's sounding boards for new ideas or strategies in the lab.

Jack had once thought that their relationship could get more serious. He knew he enjoyed spending time with

Amy and enjoyed their conversations and arguments. He found himself waking in the morning and eagerly anticipated a day in the lab with her. Jack certainly found her attractive, as did a number of other men at the institute; however, he realized he never felt that spark of romantic interest when he was around her. He recognized one evening that he enjoyed her presence more as a friend, somewhat like a sister, but he had no romantic feelings for her. He never spoke to her about his feelings, but after that evening, he never again doubted their relationship as becoming anything other than best friends.

Amy was probably the brightest up-coming scientist he had ever met. She routinely read the latest issue of at least ten journals, and in her spare time she read and outlined textbooks on immunology, neuroscience and even virology. It was not uncommon for her to back up some scientific argument by quoting the original experiments, authors and even the institution where they were completed. She was truly a scientific smorgasbord of information on subjects that were often only distantly related to her field of study. This immense knowledge led to some competitiveness between the two of them. That's not to say that Amy would steal a project or undermine Jack's experiments so that she could get ahead of him. Their competitiveness served to spur the two of them on harder in their research, causing them to work longer hours and weekends.

The competitiveness could benefit each of them too. Jack would attend seminars and read his own journals in an attempt to stay one step ahead of Amy. Earlier today, he had suggested that Amy look into using a panel of synthetic proteins against the ricin toxin in an effort to block its effect. Jack had attended a seminar from a scientist at Abbott Laboratories where the speaker had described the company's panel of over 10,000 small proteins that had been randomly constructed in an attempt to identify an antiviral protein that would affect the protease gene of HIV. Remembering that talk, Jack had suggested that Amy contact Abbott about using that panel of proteins against the ricin toxin.

Jack spent the rest of the evening riding up the coast through the small towns. Rides like these could often clear his head and allow him to think about his work and his life more clearly. The feel of the wind in his face and the constant thumping sound of the large v-twin engine under him helped to drown out outside thoughts. He had a favorite little beach up this way that was quite secluded. He had never brought anyone here except Ace.

Today there was only one woman on the beach apparently trying to get a late suntan. Jack left his bike on the side of the road and walked along the beach away from the sunbather. When he got around the bend in the cliff face he sat down and watched the waves roll in. Having grown up in Montana on the great plains in the shadow of the Rocky mountains, Jack had never had much experience with the ocean. He had seen the ocean when he was young, but he never lived near it. He never learned to swim in the ocean or surf the waves. He never fished so far out that he could no longer see the land he had come from.

Now that he lived and worked close to the sea, he realized he didn't really want to try those things. All he really wanted to do was sit on the sand and watch the waves roll in. There was something soothing and hypnotic about watching the ocean expand and contract against the land this way. To know that thousands of years ago, the Native Americans probably sat in this same spot and watched this same water.

The sun was beginning to set now. It was still quite high above the horizon, but it was at that point in the sky where if you looked away for a few minutes and then looked back you could see it had moved lower in the sky. As it sank toward the horizon of water, you could swear the sun moved more quickly to its nighttime resting spot. The clouds that always hung out over the water offshore lit up as brilliant red and orange streaks across the sky. Not unlike the sun sets back home after an evening thunderstorm.

This was possibly the best part of living so close to the ocean: unobstructed sunsets. Although the warm weather and the ability to sit on the beach in January were nice, the sunsets and their ability to clear your mind and ground you in reality were the real bonus. At least that is what Jack thought. The sun had disappeared and the light was quickly fading but Jack didn't want to leave just yet. He got up and walked a little farther along the beach and closer to the water, watching the water come closer and closer to his feet as he played a game of keep-away with the waves.

If Ace were here, he would be barking at the waves and trying to chase them back to the ocean. He loved to play in the water when he was at the lake, but the ocean waves presented a new challenge and apparently he felt that his owner needed protection from them.

Jack sighed as he noticed it had become dark. Even in Southern California the days were cut too short in the winter. Tomorrow he would do what he could to help Samantha find out about her brother. Jack headed back to his bike along the softer, drier sand that made walking twice as much effort as his feet continuously slipped back a bit for every step he took forward.

As Jack was approaching his bike, he noticed a man down the road leaning against a car. The man had a camera hung around his neck and was holding a newspaper over his face so Jack couldn't see his features clearly, but what interested Jack was that the man seemed to be reading without any light around.

Chapter 8:

The next morning Jack once again woke to Ace's warm breath on his face. "Hey buddy, good morning to you too. Okay, okay I'm up."

"Do you want to go for a car ride today?" Jack asked Ace. "I need to watch someone, so we can take the Jeep out and drive around for a while. You can keep me company."

Ace eagerly jumped into the passenger seat of the Jeep. The passenger window and side of the windshield were dotted and smeared from Ace's nose-prints during these rides. Jack packed a cooler for lunch for himself and a gallon of water for Ace and headed over to the address Samantha had given him for her brother Brandon.

Brandon lived downtown in an area that was badly in need of 'urban renewal'. Jack maneuvered the Jeep around pot holes and double parked cars. Many of the houses were boarded up and looked as though the last paint applied was during the Eisenhower administration. In many cities, large, old houses like these would be yuppified by now with prices far exceeding anything the local residents could afford. But not here yet. This neighborhood still reeked of depression and decay. Jack passed orange and pale-blue houses and a two-story brick apartment building with a faded advertisement along one whole wall that pointed the weary driver to the McHale Hotel.

Jack parked a couple doors down from Brandon's place. According to Samantha he lived on the second floor flat. It was hard to imagine that a guy could buy a beautifully restored Vincent motorcycle and yet live in what wouldn't even qualify as a half-way house. This was the kind of area that people lived in to disappear. Jack called Brandon's phone number from his cell phone to make sure he was still in the apartment. A man answered with a groggy stream of obscenities.

"Well Ace, sounds like Brandon had a rough night. Guess we just kick back here for a bit."

Jack leaned the seat back to read the morning paper he had brought with him and to drink his coffee. It was after 11:00 when Brandon finally poked his head out the front door. He sat on the front step of the house dressed in tan slacks, a shirt and a tie. He was tall and thin with ruggedly handsome looks. Brandon didn't seem to be heading for any car or even getting ready to leave. He just sat on the front step looking straight ahead. Five minutes later a Nissan 300ZX drove up to the curb, and Brandon approached it and got in. Jack thought it couldn't be another coincidence. This sure looked like the same silver 300ZX he had seen on Monday. Although there were other silver 300ZX cars in the city, this one had the same dark tinted windows and bright chrome wheels that Jack had seen the other day. He realized his jeep might have trouble keeping up with such a fast car.

Jack followed the Nissan down Broadway Street toward downtown careful to keep the silver car in sight without driving so close the driver would notice. The two cars drove through the outer edges of the downtown. This was a marginal neighborhood that was just beginning to undergo revitalization. Small stores and restaurants had begun to move in and now shared breathing room with the liquor stores, tattoo shops and auto repair places. Jack had no idea where the driver of the Nissan was heading. He maintained his concentration on the other car hoping he was doing this correctly.

The silver Nissan pulled over to the right side of the road in front of a small sandwich shop set back only about ten feet from the curb. Brandon quickly jumped out of the car and jogged up to the store. Jack drove past the car trying to find a parking spot on the same block. The driver of the Nissan had been lucky and had taken the only spot on the street. Jack reached the end of the block and stopped at the red light. Should he try to drive around the block and hope to come back to Broadway Street at the time Brandon came out, or should he drive on and park up the street and hope the Nissan continued on Broadway? Jack sat in the Jeep at the red light knowing he only had seconds to make a decision before the light turned green and the car behind him would want to move on.

Jack decided to go around the block. He quickly turned right while the light was still red and sped down the street to the next light. Ace went flying against the window with a yelp while Jack turned right again and drove down to the next intersection. Turning right at this street Jack found himself back at the intersection on Broadway. He felt in a better situation now. He was stopped at a light with a sign posted below NO TURN ON RED. Jack could see the Nissan parked on the street in front of the sandwich shop now, and no one was behind him. He could just wait here until Brandon came out of the sandwich shop. Suddenly Brandon came running out of the shop and jumped into the silver car. Jack watched in shock as the Nissan sped out of the parking spot and on through the green light on Broadway.

Jack looked around quickly, and not seeing a cop, hit the gas of the Jeep squealing the wheels around the corner. 'Smooth move Jack,' he thought. He barely made it through the yellow light on Broadway. Hecame up behind the Nissan as it was turning left onto 12th Street. Jack got back behind the Nissan and allowed his heart to slow down a bit. He opened a window to cool off and turned to Ace, "Sorry buddy, you okay?" Ace had retreated to the protection of the foot well of the passenger seat.

Jack followed the Nissan being careful to stay at least a half block back and occasionally letting a car get between the two of them. He hoped the driver of the other car wasn't watching for a tail, because he was sure he wasn't doing this right. Brandon and the mystery driver drove toward the south side of town where a lot of old warehouses had once been built. Now most of the buildings were abandoned or were used for long term storage for businesses. The main street through this area of town looked okay if you were just driving by, but if you looked more closely you could see grass and weeds growing up from the cracks in the neglected sidewalks. Many of the streetlights had long ago burned out and buildings and street signs were tagged with graffiti that bore only a passing resemblance to English. This was probably an area of town where one could find drugs in the doorways and hookers on the corners.

Jack followed the other car down the main street and watched it turn down a small alley. He knew he couldn't drive down the alley without the other driver knowing he was being followed now, so he pulled past the intersection and quickly jumped out of the Jeep, running back to the alley. He could just see the Nissan turning left onto the next street. Jack ran down the alley to where the Nissan had turned, and he saw a very small street that was probably used more by forklifts than by cars. Certainly a truck could never have driven down this street. The Nissan had parked in front of one of the warehouses, but neither Brandon, nor the mystery driver could be seen.

The warehouse where the Nissan had parked looked considerably different than the ones around it. Jack couldn't put his finger on it, but he knew that something didn't look right. Looking around a little more he saw a homeless man and woman huddled beside one set of steps. They had a little mutt of a dog with them. All three appeared to be sleeping. Should he walk up to the building and try to see Brandon? But what if they came out right then? Jack didn't think that

Brandon would recognize him but he knew this wasn't the kind of place that people just went for a stroll. Jack looked around and got an idea.

He ran back to the Jeep and opened the back door. He still had his coveralls in there from the last time he worked on Tim's car. He quickly slipped them on and found a hat that had been wedged under the back seat for probably the better part of the last decade. He then grabbed his newspaper and wrapped it around a Snapple bottle so that it would look like a 40 oz. bottle of beer. The coveralls were long enough that they covered most of his hands and they definitely covered his new sneakers. He called Ace out the back door and headed back to the small alley.

"Stay close buddy," Jack said as he took off Ace's collar. "Keep your eyes open."

Jack turned the corner onto the small street and quickly walked on. He suddenly stopped and thought for a minute. His disguise as a street bum wasn't going to work if he walked like this. Jack slumped his shoulder and turned his head down to the ground. He took a sip from the Snapple bottle splashing some of the liquid down his chin and onto his coveralls. Jack continued his walk along the street in the direction of the Nissan, however at a much slower pace.

Across the street from the warehouse, Jack slumped down beside a garbage can to watch the building. From this distance he could see what was troubling him about the view before. All the buildings around here appeared to have been built in the 50's or earlier with rusting steel walls and broken windows. The building where the Nissan was parked was obviously built much more recently. This building had a new concrete stair going up to dark tinted glass doors. The walls were concrete but had been painted to look like rusting steel walls to match the surroundings. Jack guessed he was looking at the back entrance of the building. There were no signs around and no parking lot.

Jack decided to get a closer look at the building. He slowly pulled himself up and stumbled across the street with Ace following close behind. He could hear a low hum coming from inside the building. Looking up, Jack could see several large power lines entering near the roof.

Walking around the side of the building to what he assumed would be the front entrance, he saw more of the concrete walls painted to look like rust streaked steel walls. There was no company sign nor grand entrance. In fact the location of a single loading dock made this look like the rear of the building. Jack was confused. If Brandon was in there, what was he going to tell Samantha? He had no idea what Brandon was doing and had no clear plan to find out. The only windows in this building were at least two stories up and all appeared to be fixed in place.

Jack looked around and saw Ace up on the loading dock sniffing at the door. Thinking he could see in there, Jack stepped up to the loading platform and stood even with a service door that contained a small window. Peering through the window, Jack was astonished at what he saw.

He was looking directly into a high-tech biotech laboratory. He counted at least six people in white lab coats standing at benches. The equipment was all new and state-of-the-art. He saw several biocontainment hoods, high-speed centrifuges and DNA isolation kits. Looking around he saw a sign indicating a biosafety level 3 room off to the left. A room typically used for the study of infections diseases that held no cure and only limited treatment options. What kind of biotech company would locate in the south side of town in the middle of the warehouse district? And who ever heard of a biotech company that didn't advertise itself?

Ace whimpered and pushed Jack with his nose. "What?" Jack whispered, and then he heard voices. Jack jumped down from the dock and hid around the corner from the building holding Ace back so that he wouldn't disclose their position.

"...fucked now man." Jack heard one man speaking as he came out the service door.

"I've said before, if they put it in the air, someone's gonna get fucked." Another man spoke up.

"Gotta light?" The second man asked.

"Yeah, here. So you think they'll get caught puttin' it in the air?"

"No question…" The second man had turned away from Jack, "…sick as a dog, throwin' up and shit."

"No shit! Didn't think it was that bad."

Jack tried to get a little closer so that he could hear everything they were saying. Were they talking about some biohazard agent? Something made here?

"A lotta people are gonna be hurt by this." The second man spoke again. "If I was a bettin' man, I'd put money on the Raiders. No way Doug Flutie is gonna feel good enough to play."

Jack laughed at himself. A football game? "Jack, you need to calm down a bit."

Jack crept quietly along the side of the building back toward the Nissan. He sat down against the back of the building next door to the biotech warehouse to wait for Brandon.

"I would love to get inside there, Ace."

Jack sat on the ground with his head turned to the ground watching the Nissan out of the corner of his eye. He realized he wasn't going to get into the building so maybe the best thing to do was to wait for Brandon to come back out.

It was more than two hours later that someone emerged from the glass doors of the building. The man was tall and skinny and wore a very nice suit. He was wearing sunglasses even as he left the building. His skin was dark and he had a well trimmed beard covering a very dark face. He walked with quick long strides toward the Nissan and stood next to the driver's door looking back toward the glass doors. A minute later Brandon came out walked over to the Nissan's driver.

The street was so narrow that Jack was less than 10 feet from the two men.

"What am I going to do?" Brandon asked the driver.

"That is not for me to decide. I just have to watch you." The driver spoke with an accent. It sounded vaguely British but not as a native of England. It also sounded as though English was not this man's native language.

"They've got me over a barrel now, Aded. I've done everything they've asked me to do. And now they treat me like this. I should show them."

Aded appeared shocked by this last remark. He straightened up and with a movement so fast it didn't even appear human, he grabbed Brandon by the arm, spinning him around and pinning him to the hollow hood of the Nissan with a loud boom like a drummer hitting a kettle drum.

"Listen to me Brandon," Aded barked. "These men are not playing games here and neither am I." He released Brandon who pushed away and massaged his shoulder. "I have been with you for a while on this one. I have a lot at stake, too. You will do what they say, and I will be right beside you to make sure you do. You will not screw around with me on this one. Am I clear on this point?"

"Yeah, yeah, sure. Jeez, you don't have to warn me." Brandon looked as though he had just seen the devil and turned back.

"Okay, let's head down to the beach then for our measurements, and then we'll meet those guys later tonight like they asked. You know I really don't like this." Brandon said.

Brandon and Aded got into the Nissan and headed down the narrow street to the next intersection where they turned left. Jack quickly got up and ran back to the Jeep with Ace right at his heels. As he turned the corner to where he had parked, he saw the Nissan coming right at him. The Jeep partially blocked him from the view so he dropped his head and stumbled forward leaning against the Jeep to give the impression he was throwing up. The Nissan drove right by without appearing to notice.

Jack turned the Jeep around and followed the Nissan, keeping a larger distance this time. He was pretty sure they were headed to Pacific Beach. Samantha had said that was where Brandon spent his days surfing.

Jack pulled his Jeep into a spot a half block down from the Nissan and took off the coveralls and slipped the collar back onto Ace. He got out of the Jeep and walked along the boardwalk toward the Nissan. He could see Brandon slipping into his wetsuit with his surfboard leaning against the back of the car. Aded was lifting a silver, metal briefcase onto the hood of the car. He opened the case and pulled out two small, black devices. One was the size and shape of a compass, the other looked like a palm pilot. He hung the compass-looking device around Brandon's neck and handed the palm pilot to Brandon who slipped it into a pocket on the leg of his wetsuit. Aded then got back into the Nissan and drove off leaving Brandon walking toward the surf.

Jack walked down toward the beach and made himself comfortable in the sand. He had brought his lunch from the Jeep and it was after 2:00 already. The sun was still high enough to warm the sand and the air at the beach. Jack slipped off his shirt and his shoes and relaxed on the sand in just his jeans. He watched as Brandon paddled out through the waves, ducking his head under each crashing wave so that it didn't push him back toward land.

Jack ate his sandwich and drank the rest of the bottle of Snapple he had brought out with him while Ace dug holes in the sand spraying Jack and everything within a five foot radius with the sand. Jack closed his eyes and leaned back to lie down in the sand. Samantha may have been right in thinking that Brandon had gotten himself into something bad. His conversation with Aded sounded as though he might have someone putting pressure on him. But what would Brandon have to do with a biotech company, and what company was in the warehouse district? Jack had hoped to be able to provide Samantha with a quick answer to her questions about Brandon. But so far today, Jack had only managed to unearth more questions and very few answers. Maybe he should try to follow Brandon and Aded tonight to whatever meeting Brandon had mentioned.

Jack looked up and saw that Brandon had finally paddled out to the other surfers and was sitting on his board to wait for the bigger waves to come in. Jack walked back to the Jeep and got a small pair of binoculars he kept in the back. He sat on the hood and looked out to Brandon who appeared to be talking to a few of the other surfers. Then he lay down on the surfboard and started paddling in as a wave began to build behind him. As the surfboard was being propelled forward by the energy of the wave, Brandon climbed to his hands and knees and up to a standing position. He turned the board into the wave and curved smoothly up to the peak before pulling the board back around and sliding back down the face. As the wave approached the shore and began to fall apart, Brandon leaned off to the right and dove into the face of the wave. Jack could see his head pop back up through the froth of the wave as he climbed back onto his board to paddle back out.

A few minutes later, Brandon caught another wave and rode this one toward the shore while angling to the south a bit. He paddled out again but was farther down from the other surfers this time. He paddled out beyond the point that the waves began and sat up on his board. Jack's binoculars weren't very strong so he couldn't see too clearly, but Brandon appeared to take the compass-like device from around his neck and open it up. He held it up in the air for a few seconds and then pulled it back down to look at it. He pulled out the palm pilot and typed something into it. Then he held up the device again before typing into the palm pilot again. Jack wasn't sure, but it looked as though Brandon was measuring the air for something. He might have been measuring something in the air or maybe he was just measuring the wind speed. Brandon made several more recordings and then paddled back toward the cresting waves and surfed in.

For the next hour and a half, Brandon repeated this routine of surfing a few waves and making recordings before he paddled all the way in to the beach. Brandon showered off at the outdoor shower on the beach and then stepped into the bathroom and changed back into his khakis and shirt. He then walked up to the boardwalk wall and sat with his back to the ocean while he typed into his palm pilot. At four o'clock Aded drove up in the Nissan to pick up Brandon.

Jack followed the Nissan through town back to Brandon's apartment and watched as Aded drove off and Brandon walked up to his flat. It was strange that a man that surfed wouldn't take his surfboard with him. Jack had never surfed himself, but the surfers he knew would never leave their boards in someone else's car all the time. Besides, it must be a tight fit in that car with the surfboard. Jack parked again and watched Brandon's front door.

Jack poured some water for Ace and decided to wait a while and see if anything else happened. Although it was nice to sit on the beach for a few hours today, he would gladly have spent the day at work in the lab instead of following someone. He just hoped this would help put Samantha at ease. Maybe he could cook her dinner this weekend and take her mind off of taking care of her brother.

The streetlights flickered on, shining their floodlights of fluorescent daylight onto the street from high atop their poles providing the residents of the apartment buildings with small circles of security in a neighborhood where security is purchased in the form of an extra deadbolt for the door or a club for the family car. Jack sighed and looked over at Ace curled up into a ball on the passenger seat. What loyal creatures dogs were, and all they asked for in return was to be part of the pack.

Jack relaxed in the driver's seat of the Jeep until shortly after 6:00 when Brandon came down the steps. He had changed into blue jeans and a short sleeve polo shirt. He walked down the street a few doors and got into an old beat up Ford Escort. Jack was glad to see he was finally going to be following a car he could keep up with. Brandon drove through the downtown and North along the beach right to Jack's apartment complex.

Brandon was going to see his sister, and with that Jack realized where he had seen Brandon before. He was the man who often came over to Samantha's apartment in the evening. She wasn't dating someone else, or at least no one that Jack knew about. Jack suddenly felt very good about his day's work. He was sure Samantha would appreciate that he had helped her look in on her brother, and he might finally have the opportunity to ask her out on a real date.

Chapter 9:

Jack showered and changed before taking Ace for his evening walk. Returning to his apartment, his thoughts were of Samantha when suddenly she and Brandon stepped out of her apartment.

"Jack, oh hi. I want to introduce my brother, Brandon." Samantha winked at Jack.

The smile on her face stopped Jack in his tracks and almost melted him right there. She was wearing a knee-length simple black dress, but she looked flawless in it.

"Hi, Samantha, and hi Brandon. It's nice to meet you."

"Nice to meet you too, Jack. I must have seen you around here before, you look familiar."

Samantha quickly interrupted. "Jack, I was telling Brandon about our dinner at the Thai restaurant and the owner named Tim."

"We could go to dinner tonight if you would like. How about it, Brandon?" Jack asked.

"Oh, I would love to, but I'm meeting an old friend of mine who is in town only for tonight. Maybe a rain check?"

"Sure, maybe another time." Jack said. "Samantha, could I maybe persuade you to join me for dinner? I could cook something."

Samantha flushed slightly, but Jack caught it. "I would love to."

"I'll cook up my famous chicken á la raspberry-mustard. Why don't you come over around 8:00? And Brandon, nice to meet you."

Jack walked into his apartment with Ace and looked around. His place was a mess with clothes on the sofa, dishes on the counter and dog toys strewn across the floor. His father would have said the place looked like it had been hit by a tornado. He spent the next half hour cleaning the place, which meant throwing everything into the closets and quickly sweeping the floor. He used Windex on the really dirty spots of the floor to avoid having to mop. Then he opened his refrigerator. It was nearly empty, and he didn't even have a famous raspberry-mustard sauce.

"Okay, quick…think, Jack!"

Jack walked down to the local grocery store and bought a few chicken breasts, a bottle of mustard and some fresh raspberries. For a side dish he picked up some green beans. As an after thought, he picked up a bottle of Zinfandel. Arriving back at the apartment, he put the mustard and raspberries in a blender along with a little olive oil and blended it all together. Then he poured it over the chicken to marinate for a while. He still had a half hour so he set the table and lit some candles. Then he put the chicken and marinade in the oven to bake for a while. Just as he was opening the wine, he heard a knock at the door.

Samantha was waiting on the front step. Fortunately, she had kept the black dress on. He was really going to have to watch himself around her. He would probably agree to follow her anywhere right now.

"Come on in. And don't mind the killer guard dog." Jack teased.

At the sound of the knock on the door, Ace had retreated to the back bedroom to find a toy to bring to the new guest. He was currently shoving an old ball at Samantha's feet.

"Oh Jack, dinner smells wonderful. So where does this famous recipe come from. I mean, is it Italian or German?"

"Let's just say it comes from Montana. My dad's idea of a 'famous recipe' was anything he could make up on the spur of the moment. I'd show you around, but my place is the same as yours. You're standing in the living room slash dining room slash entertainment room. And over there is the kitchen. That door is the bathroom. Down the hall is my bedroom. There, tour done. Now then, the chicken will be done in about 10 minutes. Would you like some wine?"

"Relax Jack. Yes, I would like some wine." Samantha threw the ball down the hall for Ace and watched as he promptly chased after it.

"You'd think he gets no attention the way he pesters guests."

"So Jack, I have to know. Did you find out what my brother is doing here in town? Does he really just surf all day?"

Jack sighed and took a sip of wine, placing the glass slowly back down on the table. "I wish I could give you some answers. I did follow him around all day today, until he came over to your place this evening. I really can't say what he does all day." Jack paused. "Let me get dinner and I will explain all of it to you after we eat. Okay?"

Jack served up the raspberry-mustard chicken and a side of steamed green beans. "So, Samantha, the other day I didn't get a chance to hear what you do for work."

"Well, I work for the Union-Times newspaper as a researcher. Not a researcher like you, I just gather information for stories, verify facts, that sort of thing. Mostly, I get my information off the internet, but sometimes I have to go out into the field to get information. I'm not the one that interviews or even writes the stories. But when you read a feature story about a land developer building a new subdivision in an environmentally protected area, I'm the one that provided all the facts on the environment, the developer and the local ordinances. It's not nearly as exciting as your job."

"I think you may have the wrong impression of my work." Jack responded.

"I got my degree in English literature from Penn State University. I always wanted to be a writer, but that doesn't exactly pay the bills. So I found this job at the paper. I'm hoping this is only temporary. I spend most nights at home writing for myself. I'm working on a novel right now."

"Really, that's fascinating. What's it about?"

"Well, I don't want to talk too much about it until it is done, but it is a crime novel about a newspaper writer that is sent information through the mail about a crime that is about to happen. The lead character disregards the letter, and then the next day an article in the paper about a homicide catches her eye. She digs a little deeper and finds details about the murder that were in the letter she received. From there, she helps the police try to solve the crime, but also tries to solve it herself so she can get the big article in the paper. I don't know, writing a novel is such a daunting task. In school I wrote short stories and had a few published in the student paper. But this seems overwhelming at times."

"Wow, the story sounds great! I know how large a project can seem at first. When I come up with a proposal for a research project, it will usually contain some grand goals that seem almost unattainable. And each day I have a setback, it seems like I will never reach those goals. It really makes me question being in research sometimes."

"Exactly. I guess you do know how I feel. I sometimes spend hours at night staring at that computer screen and nothing comes into my head. It is just a complete blank. And I wonder if I can ever complete this. And even if I do, will it be good enough to sell? I try to tell myself that I am only writing this to prove to myself that I can do it, but I also really want to sell this novel and start writing full time.

"I've never told anyone about writing this book. I've been thinking that if I didn't tell anyone and I failed at it, no one would know except me. I wouldn't feel bad about showing my face at family meals or at the paper. And no one would be asking me how the writing is going or if I got published yet. I really want to succeed at this, but I'm afraid that I just don't know what I am doing."

"I wish I could help. But I suppose this is something that has to come from within you. If you ever want to bounce ideas off me, just let me know."

They had finished their meals and Jack cleared the table. He realized that he didn't have anything for dessert. He looked in the freezer and found a quart of vanilla ice cream. He scooped out two bowls and covered it with some of the fresh raspberries he had just bought with a little chocolate sauce he had in the cupboard.

"The meal was great. Thanks, Jack. I needed to take a break from writing tonight anyway."

Jack poured the last of the wine into the two glasses. "Okay, I'll tell you about today."

Jack told Samantha about following Brandon and Aded to the warehouse district and about the strange biotech company. He told her about watching Brandon surf and take measurements while he was on the surfboard. None of this seemed to be putting her at ease.

"Oh, one other thing, when he and this guy named Aded were talking about meeting some guys tonight, it didn't sound like Brandon wanted to go, but Aded was threatening him. I got the impression they knew each other well enough but Aded didn't seem like a guy I would trust."

"Oh, this is awful," Samantha shook her head. "What has he gotten himself into? You don't know who these guys are?"

"No, but maybe you could find out about this company. You said you research this kind of thing. They must have a deed to the property and some sort of business name filed with the city. And if they have a BSL3 facility, they must…"

"A what? A BSL3 facility, what's that?"

"Bio-Safety Level 3 containment. There are four levels of bio-safety containment with 4 being the most restrictive. A level 4 facility is only found in a few places in the world with only two here in the U.S. A level 3 facility is used to study organisms that are known to cause life threatening diseases without effective treatments like HIV or anthrax bacteria. In order to construct a level 3 facility, you need to file permits and have the Department of Health inspect the facility on an annual basis. You should be able to find those permits on record."

"So, what does this company work on at their BSL3 facility?"

"I don't know. I couldn't see enough of what went on inside the building. But hopefully it would be listed on the permit."

Jack cleared the remaining dishes off the table and fed Ace his dinner. Samantha sat in her seat absently picking at a small ruby studded ring she wore on her right hand.

"What should I do? I don't want him getting hurt."

"I don't understand Samantha. It seems like you two are pretty close. He comes over to your place quite often. Why don't you just ask him?"

"I've tried, but he just tells me it isn't my concern. He told me he could take care of himself. You don't understand. When he was in college, he got into some trouble. All growing up he got into trouble. The police would knock on our door at 1:00 AM to tell my parents that he had been caught spray painting signs, or the school would suspend him for smoking in the parking lot. Mom and Dad were always trying to bail him out of trouble. Then in college, he was arrested for possession of marijuana. He asked my parents for help, but they refused. They thought that if he had to deal with the arrest himself, he would start to grow up. Ever since then he won't even speak to them. I helped him out then and I have ever since. If I push him too hard on this, he may leave me too, and without me, he has no family. I vowed always to take care of my little brother. I want to make sure he isn't in trouble now, but I can't force him to tell me about it.

"I think we should find out who he is meeting tonight. I mean, it must be important. He already lied to you about what he was doing when you asked him to dinner."

"Whoa, I don't know. We don't even know where or when he is meeting these people. And besides, it's night. It'll probably be really hard to follow him without his knowledge." Jack could see the hurt expression on Samantha's face, and based on what she had just said, he had the impression that she would do this with or without him. "Okay, I'll help you. Why don't you give him a call to see if he is even still at home and then we can make a plan?"

Chapter 10:

Brandon was at home and was continuing to use the excuse that he was meeting a friend later. Jack pulled the Jeep around and Samantha got in. They drove over to Brandon's apartment and waited down the street for him to leave.

It was after 11:00 when the Nissan 300ZX pulled up. Jack pointed out the car to Samantha as Brandon came down the steps and got in. They followed the silver car back uptown along the coast for a while passing out of the city and through several coastal towns until they arrived at a small town named Santa Rosa. Jack was able to keep some distance from the Nissan on this long straight road that was still remarkably busy even at this time of night. Brandon and the driver drove into town before parking outside a small office building.

The building lobby where Brandon and Aded entered was dark. In fact the only light in the building came from the back of the building on the lower floor.

"Should we try to follow him inside?" Samantha asked.

"I don't think we should. The lobby is dark and the building looks closed. They would surely see us. Maybe we can sneak around to where the light is on. Perhaps we can see something there."

The front door listed three law firms and a chiropractic practice as located on the first floor. Jack didn't think they were headed for a chiropractic appointment. The windows of the building were located at shoulder height and several trees grew near the window with the light on so Jack and Samantha were able to hide in the shadows. Peering over the ledge, they could both see Brandon and Aded sitting on one side of a large desk. On the other side of the desk was one man seated in a large high back chair and another man standing next to him. The man in the chair had dark skin and a short, well-trimmed beard similar to Aded's. The man standing was very tall, white with long blonde hair pulled back into a pony tail. As he shifted on his feet, a bulge that Jack assumed was a gun, could be seen in the small of his back.

The window had one large pane of solid glass at the top and a smaller pane of glass at the bottom that slid open for a screen. This window was open so Jack and Samantha could overhear the conversation inside the room. The seated man was speaking.

"...have been pleased with your cooperation thus far. The wind measurements will be useful for our plan. I called you here personally to assign you..."

Brandon interrupted, "Mr. Saresh, I think..."

"You do not interrupt me!" Saresh slammed his fist onto the desk. "And I have told you not to use my name again. You will not think about this, I will tell you what to do. Do I need to remind you that you are deeply involved with us now?"

Brandon flushed deep red and began to get up, but Aded clamped his hand around Brandon's wrist so hard that Jack could see him wince. Brandon sank back to his seat.

"I understand sir, I just don't see how I can be of any more help to you. I've done everything you…"

Saresh rose from his seat now and raised his voice even louder, "Does 'everything' include stealing money from me? You steal money from me after all I have given you?" Saresh reached behind the blonde man and grabbed his gun. He pointed it directly at Brandon's face. "You will do what I ask of you, and you will not argue with me again. Your future is very much in our hands. I will not hesitate to pull the trigger if I think you will betray me again."

Brandon's arms were folded in his lap with his hands clasped tightly, but that did not stop them from shaking. "Yes sir, I understand."

Saresh lowered the gun onto the desktop and returned to his seat. Throughout all of this Aded and the blonde man never flinched.

"Good. Now then, I was beginning to say that I have a new assignment for you. You are very valuable to us as an American citizen." Saresh reached into his top drawer and pulled out an envelope, pushing it across the desk toward Brandon. "In this folder is a roundtrip plane ticket to Seattle. You fly there on Friday as though you are going for business. You will also find a rental car reservation. You will drive to Vancouver, Canada on Saturday and check into the King George Hotel. A man will contact you there and give you a case to bring back to me. Coming back across the border you will not be questioned because you are American. You will return to me with the package on Sunday."

"Yes sir."

"The case will be locked so you won't be tempted to open it and steal anything more from me."

"Yes sir."

Brandon reached for the envelope and slipped it into his jacket pocket.

"I know you have that motorcycle at your sister's apartment, Brandon."

Brandon looked up at Saresh with a shocked look on his face. "Uhh, umm"

"Do not lie to me. We know you stole the money and we know what you did with it. You will repay us at a future date. Now leave before I get angry with you."

Aded rose and pulled Brandon up by his arm leading him to the door.

"And Brandon, I don't have to remind you what will happen if you don't return on Sunday with my package."

Jack pulled Samantha's arm away from the window and led her away from the building. "I think we need to talk to Brandon now. He's obviously gotten himself involved in something he can't control anymore. We need to convince him to go to the police."

Samantha merely nodded; she had tears streaming down her face. Jack pulled her into his arms and held her under the moonlight. He had wanted to hold her again since that night at Tim's place, but he had hoped the next time it wouldn't be just to sooth her tears. Her shoulders shook against Jack's chest and her legs seemed to shake. Finally she collapsed to the ground.

"I just, I just don't know what to do. I'm afraid."

Jack cradled her on the ground. Maybe if he gave her a little time she would be able to walk back to the Jeep. "Maybe it's not that bad. He probably just got involved in something over his head. I bet the police can help him out of this. Why don't we head back to town and call him over tonight."

Samantha merely nodded her head. She allowed Jack to help her to her feet and back to the Jeep. The drive home was quiet as Samantha just stared out the passenger window.

It was after 1:00 when they arrived back at Samantha's apartment. Jack noticed that she wouldn't even turn her head toward the motorcycle. She had replaced the sheet since his visit on Monday night. She immediately went to the phone and tried calling Brandon but there was no answer. It had already been a long day for Jack and he was tired, but he didn't want to leave Samantha like this. He decided to wait with her until she got a call back from Brandon.

Not knowing what to do or say, Jack just sat on the couch and held Samantha's hand while they waited. The last time Jack looked at the clock it was after 3:00. He woke up at 6:30 the next morning with Samantha's head on his shoulders. Brandon must have decided it was too late to call that night. If only he had known what that would do to his sister.

Chapter 11:

Jack woke Samantha and explained that he had to get into work that day but that she could call him at any time. She said she would go to work too. Brandon would likely call her there and she could arrange a meeting for the three of them that night.

Jack didn't function well on so little sleep, but the ride in to work the early morning had helped to clear his head. He was pleased to find that he had arrived before Amy. That would start his day out nicely. Since it was early and no one was in yet, Jack spent the first hour completing work.

Unfortunately, since most of the equipment currently used in laboratories is quite expensive, only one or two pieces of each instrument can be purchased for use by a whole laboratory creating delays in experiments due to the backlog of users. That was one of the reasons Jack liked coming in so early. He could have access to the machines before anyone else. The other reason was the extreme silence in the laboratory. True, a lab was never completely silent, as machines hummed, air jets turned on and off and exhaust fans blew air around, but that all became background noise. Early in the morning, there was no talking, no radio and no other foot traffic. By the time others came into the lab, Jack had started all his experiments for the day and could afford a coffee break, which on this day was badly needed.

"Hi Jack, I knew you'd be in before me today." Amy was already standing at the coffee maker waiting for it to finish brewing a pot. "I was here late last night and I overslept this morning. I thought you might come in last night."

"Sorry, I got held up with something that took a lot longer than I thought."

"Do you want to talk about it?" Amy could see the dark circles under Jack's eyes from lack of sleep.

"No!" Jack said more tersely than he had intended. "But thanks for taking care of that work for me yesterday." Jack tried to soften his tone, "I think this experiment is going to work this time. I'm just running the results through the machine now."

Jack took his coffee and headed down to the den with several journal articles. The last thing he wanted to do right now was deal with a lot of people in the lab. Part of him was just dead tired from the previous day, but that wasn't all of it. Part of him ached for Samantha. It had only been a few days since he had first gotten to know her, but he couldn't get her out of his thoughts. He wanted to call her right now to see how she was doing, to see if she had heard from Brandon, to comfort her in any way he could. The coffee didn't work and before Jack knew it he was asleep on the sofa.

It was after noon when Jack woke up. The nap had been good for him, and he felt a lot better now. And besides, the experiment he started this morning would probably be done about now. Jack walked back to the lab and ran into Jason.

"Ahh, here you are. I was hoping I could hide out down here, but I guess I won't be alone now"

"No, no, I was just heading back up to the lab." Jack looked at the down-turned eyes and slumped shoulders of Jason. "Is something wrong?"

"I've been working on this project for Dave for three months now, but I can't get anything to work. I just don't know if I should continue here. I'm not sure I should continue in science."

Jason was a technician in the lab and worked under Dave on a project identifying cellular protein involved in aging. Actually Jason wasn't just any technician; he was the most necessary person in the lab after Dr. O'Leary. He had been in the lab for six years, which was a very long time for a lab like this. Most technicians in a non-profit research facility or at a university were paid considerably less than their counterparts in industry. The low pay and the constant turnover of short term employed post-doc researchers caused many technicians to leave for the stability of an industry job. Because Dr. O'Leary considered Jason such a vital asset to the lab, he was well paid for a technician at the Hoffman Institute, however he had talked recently about going back to school to get his Ph.D.

"Jason, everyone has these problems in science."

"The problem is, I don't know what the problem is or where to start looking for it. I am doing the same experiments I have done for the past six months, but suddenly they just stopped working. I don't know what is wrong."

"You're probably just in a slump. We all know how it is when nothing works, not even the simple experiments, and then just when you are ready to give up and call that phone number on TV about truck driving school, BAM!" Jack clapped his hands together. "Something breaks and everything works for the next several months. You just have to wait it out for that break. Have you talked to Dave about this?"

"No, he doesn't want to hear about this. He just wants results for his paper. I can't get these cells to grow. We have been trying to grow these skin cells from the old mice, but they keep dying in the lab. I've tried everything."

"I've got an idea. I read about this new serum formula they were using to grow skin cells. I have the paper on my desk because there was another technique I was trying to use. Come back with me and I'll dig it up for you."

"Really?" Jason brightened up, "Thanks, Jack. I don't know why I didn't ask you earlier. You always seem to have something new to try."

Jack led Jason back to his desk and gave him the paper. On his desk was a message to call Samantha. Jack tried the number but only got her voicemail. Now he knew his day was going to be shot. He had to find out what she had called about, but he had no way of reaching her now.

Jack walked across the lab to check on his cells. Yesterday, at Dr. O'Leary's suggestion, Amy had added the new anti-viral protein, fused to the transporter peptide, to his cells. Then she added a sample of Adenovirus to the cells. Now he placed the cells under the microscope to examine them. Jack saw that all the cells looked healthy in the dish that had received the anti-viral protein but those that grew in a dish that did not receive the protein appeared sick and dying. This was great news! Jack finally had a break in this project. His approach to blocking the virus appeared to be working.

What Jack had said to Jason about being in a slump was very much true of most scientists, and Jack had just been in a slump the past few months. Dr. O'Leary had been on his back to get this anti-viral protein working. In his first few months in the lab he had identified the protein and had shown that it was very effective at blocking the Adenovirus, but then he had hit a wall. Today was the first big break he had had in a long time. He was still a long way from solving all the problems in the project. He would need to repeat the experiment of today to verify his results. And then he would have to show that it would work in a whole animal. But today's experiment, Amy's help yesterday, was a big breakthrough.

Just then Amy walked past his bench with her head buried in a journal. "Amy, Je t'aime." Jack grabbed Amy around the shoulders and kissed her on the forehead. "It worked, the protein blocked the virus! Thank you, thank you!"

Amy looked flushed, but slowly a smile spread across her face, "I'm so happy for you! Maybe we should celebrate. We should go out for margaritas tonight to celebrate."

"Yeah, oh wait. No, I really can't tonight." Amy looked saddened but tried to keep the smile on her face. "But I promise we will. I owe you for this one. How about a rain check?"

"Okay, some other time. Besides, I'm probably going to be here late tonight getting work done. Some other time, Jack."

Jack realized he was too tired to think straight in the lab that day so it was better to call it a day and start again tomorrow. He grabbed his jacket and bag to head out to his bike. On the way out he saw Jason again and slapped him on the back. "It worked, Jason! See, you can get out of a slump. It just takes time."

Jack didn't give Jason time to respond as he ran for the door. It was only 3:00 and he knew Samantha probably wouldn't be home until later, so Jack rode home and picked up Ace and the Jeep and headed out to the beach to kill some time.

Chapter 12:

Jack slipped the leash over Ace and walked along the boardwalk. The water was about 100 yards out to his right, and small shops selling everything from sunglasses to t-shirts to beer lined his left. Up ahead about a mile away, he could see the outlines of the roller coaster that flew its occupants out over the ocean and back to land.

There was still something bothering him about the biotech company he had discovered the day before. Although it was not unusual for a company not to advertise its upcoming product, all biotech companies he knew advertised their name and general product focus. The capital investment required to bring a new drug or product to the market was so great that these companies required constant name exposure and news coverage to raise funds from wealthy investors. The company Jack found in the warehouse district didn't seem to be too concerned with advertising itself which would lead one to guess that the company was well funded already.

Ace pushed his nose into Jack's pocket causing him to stop.

"What? You think I have something in there for you?" Jack reached into his pocket and pulled out a tennis ball. "Well, look what I have."

He reached down and unhooked Ace's leash releasing the dog to dance around in anticipation. Jack threw the ball out over the sand and watched as Ace raced after it. The dog pounced on the ball creating a cloud of sand dust and a small crater in the sand that looked like a 80-pound bomb had just landed there. Jack watched as the ocean breeze slowly dissipated the dust cloud. Ace came running back and dropped the ball at Jack's feet to repeat the game.

When Jack arrived back home, he found Samantha sitting on her front step. Her hair was in disarray and her eyes were puffy from lack of sleep. She looked like a lost puppy waiting for her owner to come home.

"Jack, I tried calling you at work after I got your message, but they said you had left already. I got home about a half hour ago to wait for Brandon to come over. I was hoping you would come back before then so you could be there with me."

"Sorry I missed your call. Why don't I brew up a pot of coffee and then I'll come over."

Jack fed Ace his dinner and then poured two cups of coffee, adding a shot of Khalua in each and sprinkling a little cinnamon over the top. Coffee was one of the few things he did know how to make well. He knocked on Samantha's door and handed her a cup.

"I do need this, thanks. Wow! This is great. You're a great cook, Jack. Dinner last night and now this coffee." Samantha took her coffee to the sofa and motioned for Jack to sit down. "Brandon said he would be over around six."

"Do you really think I should be here? I mean if you want me to, I will, but I was thinking that he might take offense to being ganged up on."

"You know as much about this as I do. And if it weren't for you, I wouldn't even know he is in trouble. I think it might help to have you here. Besides, I really want you here."

Jack drank his coffee in silence while he tried to envision how the conversation might go. He was afraid that Brandon would get mad at being accused by his sister especially with a stranger in the room. Jack didn't want to be seen as the reason that Brandon wasn't speaking to his sister.

Brandon knocked on the door and walked right in. Only a family member would think of doing something like that. Jack could tell these two were very close.

"Oh, sorry Sam, I didn't know you had company. Oh, hi Jack."

"Brandon, we were just sitting here waiting for you," Samantha said.

Brandon walked to the refrigerator to get a beer. He poured it into a pint glass as he walked back to the living room to sit in a large chair that sat at a right angle to the sofa.

"Jack, I love your bike. Those old Harleys really rock. What year is it?"

Maybe this wouldn't be so bad after all, Jack thought. "It's a '60 Panhead."

"Wow, a Panhead. I don't think I've ever seen one up close. Did you buy it in that condition?"

"No, I spent about a year restoring it. It was in pretty good shape when I bought it, but I was in graduate school so I only had time on the weekends to work on it. It's not a complete restoration, but it's very close to original. It may not be as comfortable or smooth running as some of the newer bikes, but there is something about the simplicity of the older bikes that I like." Jack didn't know if he should bring up the Vincent motorcycle under the sheet in the hallway.

"What's a Panhead?" Samantha asked.

Jack looked at Brandon to answer the question but Brandon was waiting for Jack. "The Panhead Harley Davidson motorcycles were made in the late 50's and early 60's. The nickname Panhead comes from the look of the cylinder head that resembled an upside down frying pan. Other Harley engines had nicknames too, like Shovelhead or Flathead. That's how owners usually refer to their style of Harley."

"Wow, you sure know a lot about Harleys" Brandon said.

"Well, when you spend a year restoring one, you can't help but learn some about them. Besides, I spent a lot of time at parts shops or swap meets finding parts for such an old bike. I learned a lot from other people."

A lull descended on the conversation while Brandon drank his beer and Jack recalled those long days trying to find parts for his old bike.

Samantha broke the quiet, "Umm, Jack, so would you like a drink?"

Samantha got a couple of beers, and brought them back. Jack could tell she was stalling and didn't want to confront Brandon. He really didn't want to stay on the topic of motorcycles, because he wanted Samantha to bring up Brandon's troubles first. Finally Samantha spoke up.

"Brandon, I asked you to come over because I wanted to talk to you about the motorcycle you dropped off here."

Brandon gave her a stern look before turning to Jack. "You want to see it? I got a 1959 Vincent. They call it the Black Lightning, although I don't know why."

Brandon started to get up, but Samantha stopped him. "We need to talk about how you got the motorcycle Brandon. I think you're in trouble. I know you couldn't afford to buy that motorcycle working at the surf shop. And besides, I don't think you are even working there anymore."

Brandon looked at Jack and then back at Samantha. "What is this? Some sort of intervention? I'm not doing drugs or selling them! I just got a new job that pays better."

"What kind of job Brandon? It's just that I'm worried about you."

"Don't start acting like Mom and Dad, sis. I can take care of myself. I know what I'm…"

Jack jumped in, "Brandon, Samantha is very worried about you. She told me you and your parents haven't seen eye to eye in the past, and she didn't even want to talk to you about this because she was worried you would shut her out too. She spent last night crying over you. She's worried about you that's all."

Brandon looked at Samantha who had tears forming in her eyes already. He turned his head down and spoke more softly, "I can't tell you about it, Sam. I really can't. I am in trouble this time, but I can get out of it. I just have to do one more thing and then I can get out of…"

"Brandon, you have to tell me what it is."

"I don't want to get you involved. You don't know these guys like I do."

"I followed you last night. I wanted to see…"

"What!" Brandon jumped up. "You shouldn't have done that. I told you I can take care of this myself."

"Brandon, please. I just want to help."

"Stop playing big sis. I can take care of myself." Brandon paced back and forth in the living room.

Jack felt it was time to play some of their cards. Letting Brandon know how much they knew might give him the opportunity to let his guard down without feeling like he was endangering his sister.

"Brandon, I followed you all day yesterday." Brandon looked shocked and sat back down on the sofa. "I did it because your sister is really worried about you. I told her everything I saw during the day and she insisted we follow you last night too. We know you are in trouble with these guys. We want to help you, get you to go to the police. They can't threaten you like that."

Samantha got up to get some Kleenex for herself while Brandon played with a class ring he was wearing on his left hand. There was an awkward silence in the room while Brandon wrestled with how much he wanted to tell them.

"They told me not to tell anyone. They showed me pictures of you coming out of your apartment and of Mom and Dad's house. They didn't say anything about them, but I know they were threatening that if I told anyone, I would be putting you in danger. The police can't help here. I've heard stories of another person that tried to turn the police on them. He died mysteriously in jail, but not before he heard what they did to his family. I don't want to put you in that danger, Sam."

Samantha just stared at her brother, so Jack spoke up, "Okay, so maybe we don't go to the police about this. But maybe you could just tell us what is going on. We won't have to tell anyone." Brandon made no move to speak, so Jack played another of their cards in an attempt to open him up. "What goes on at that biotech company in the warehouse district?"

Brandon looked with wide eyes and an open mouth that betrayed his attempts to appear calm. He got up and went to the kitchen to get himself another beer. When he sat back down on the sofa, he started his story.

"When I first moved out here, Sam, I was working at a surf shop and just surfing for fun in my spare time. But it costs so much to live here. I was living in that neighborhood, and I couldn't afford a nice car or even a motorcycle. I hardly had enough money to even go out at night. Then about six months ago, this guy comes up to me on the beach as I was coming back in. He asks me about surfing and I'm thinking he might want some lessons, so I talked to him for a while. He tells me that he is running a study and needs to take wind speed and direction measurements off shore on a daily basis. He offered to pay me to take this wind speed instrument out surfing. He wanted me to measure the speed and direction of the wind and record it in the waterproof palm pilot he gave me. He was offering a lot of money and I figured I was already surfing, so why not?

"Then about 3 months ago, after I had been doing this every day, he tells me he needs a favor. He wanted me to go up to Santa Rosa and pick up a package for him. I figured he had been paying me so good for just measuring the wind that I could do a favor for him. So I went up to Santa Rosa and that is where I met Aded. He had the package and insisted on coming back to town with it. After that Aded would always work with me. The other guy, who told me his name was Carlos, I never heard from again. He's the one I heard went to the police.

"Aded had me doing more and more things with him. And he started paying me every week more than twice what I was getting paid to just take the wind measurements. I quit the surf shop job and just worked with Aded and took wind measurements while I was surfing. I didn't know what I was getting into. About a month ago, Aded took me downtown to the warehouse district to meet some guys and that is when I first saw that company. I guess it is a biotech company, but they told me it was a drug company. I didn't really care as long as they kept paying so well. I was going to start sending money home to Mom and Dad to pay them back for helping me out so much in school. At least before the pot bust.

"Anyway, I started wondering what kind of research project they were doing that required measuring the wind speed all the time, so I asked some questions. That's when I found out how deep the shit I was in. Aded took me up to Santa Rosa to meet Saresh. He told me about a secret project called Windswept. The company downtown has developed a cure for the flu, but the FDA won't approve it yet. They say it can take years for the FDA to approve a new drug even if they have shown it works and it cures something. So they have this plan to spray the flu virus into the wind from offshore and get people in the city sick. Then they are going to give everybody their new drug. They say it will make millions of dollars and the stock value of the company will go way up. They gave me a bunch of stock so I can make money from it too."

Brandon rose from his chair and walked over to Vincent motorcycle in the hallway. He ran his fingers along the gas tank shrouded by the sheet.

"I don't know that much about drugs or viruses, but it still didn't seem right to get a bunch of people sick with the flu just to prove a new drug works. Anyway, I tried to get out then. I told them they were demented, and I was going to the police. That was when Blondie, this tall scary looking blonde guy that is always with Saresh, grabbed me around the throat and threw me against the wall. He beat me up pretty good, and then they showed me pictures of Mom and Dad's house and your apartment. You were in the pictures, Sam. They let me know what would happen if I tried to go to the police. They said they had a lot of money riding on this. I didn't know what to do. I just kept working for them, and then I heard rumors about that other guy and how he had tried to go to the police. That really scared me.

"I haven't had to do too much for them lately. I keep making the wind measurements while I surf and I run some errands for them. They have been paying me pretty good and I will get a lot of money when the company does well in the stock market."

Jack stood up and slowly walked to the kitchen and then back to the living room. Finally he went back to the kitchen and got a beer for himself and one for Samantha.

"I can't believe that after they beat you up and threatened you like that, you're still working for them." Samantha said.

"Sam, if it was just me, I would take my chances and leave and go to the cops. But I'm not going to put your life on the line or Mom and Dad's."

"Brandon," Jack spoke up, "What's the name of that company?"

"Intech."

"I'm going to see what I can find out about them tomorrow at work. I'll try to see what I can find on this drug of theirs. Although if it's not approved, they may not have published their results yet." Jack was thinking out loud. "Well, I'll see what I can find."

"Brandon, I can't believe you'd get mixed up with guys like these. And what about this trip to Seattle this weekend? What's that about?"

Brandon again looked like he had been blindsided. "So you heard that too? Ok, so I have one more thing to do for them. But I think this is the last one. I have to go to Vancouver to pick up some package for Saresh."

"Do you have any idea what they want you to get in Vancouver?" Jack asked.

"No, they usually don't tell me. I just have to pick up these briefcases from people. Sometimes I have to go to Mexico to meet people and get these cases. They say because I am American, I can get past the border patrol with fewer questions."

"I don't know. Something doesn't sound right here." Jack decided not to ask Brandon about stealing money from Saresh. Samantha didn't look like she could take too much more. His guess was that these cases were full of money and Brandon was lying about not knowing what was in them. There still were a lot of questions, but Brandon had provided at least some of the answers.

"Brandon, I think what Intech is planning to do is a lot more serious than just getting a lot of people sick with the flu to test their new drug. The flu virus is one of the biggest killers on Earth." Jack sat back down on the sofa next to Samantha.

"The flu, no way. I've had the flu lots of time. It doesn't kill you."

"For the average healthy adult, the flu is not normally life threatening. It will just make you very sick for about two weeks. But even the run of the mill influenza virus is seriously life threatening to the very young and old and to people who have a compromised immune system such as bone marrow transplant recipients or HIV patients. And that's just the average influenza virus that goes around every year. If it's a strain of the virus that people have not been sick with before, it can kill even healthy adults."

"No way. I've never heard that before."

"In 1918, the influenza pandemic that circled the globe killed 19 million people. That's more than all of World War I. This could really be serious."

"But they told me they had a cure for it now. They even showed it worked in animals and it was safe for people."

"Drug design can be very tricky. A lot of times what works in animals, never works at all in humans. The immune system can work very differently, or the human body could break down the drug to an ineffective compound that the animal body couldn't do. That is why the FDA requires small controlled studies over a period of several years to see if a new drug is truly safe and does what it is supposed to do. Even then, many drugs get recalled after they have been sold for a year or two, and the FDA finds out some long-term effects. Even if this drug is a miracle cure, how is Intech going to get it out to everybody they infect with the virus in time to stop the disease. This really doesn't sound good, Brandon."

"They never told me all that." Brandon buried his head into his hands. This whole time Samantha just looked from Jack to Brandon without saying anything. Jack looked at his watch. Somehow the time had really slipped by, it was midnight already. Brandon saw him check his watch.

"Brandon," Samantha said, "Whatever happens, know that I love you. I'll help you with this and we'll get out of this thing."

"Sam, not this time. You've always put yourself on the line for me. I got myself into this and I will get myself out. I promise I'll make it up to you."

"Brandon, I think this may be more than you can handle on your own. I think if we all put our heads together, we can get you out of this. And then you can make it up to your sister. But right now, I think we could all use some sleep."

"I need to fly to Seattle tomorrow. I'll call you when I get back on Sunday Sam." Samantha stood and gave Brandon a hug at the door before watching him leave.

"Jack, I'm sorry I got you into this." Jack hugged Samantha as tight as he had ever held any woman in his life. He wanted so much to take the fear and sadness out of her. He wanted to know her happy and carefree. He wanted to know the Samantha he saw on the front walk just three days ago.

"I can stay here on the couch if you want."

"No, that won't be necessary. Just knowing you are next door is enough. And besides," Samantha smiled, "Ace will worry if you don't come home two nights in a row."

It was good to see she still had her sense of humor. "Call me whenever you want. I'll look up everything on Intech tomorrow and you see what you can find in the city records. We'll help get Brandon out of this. Don't worry."

Jack knew he was going to have trouble falling asleep that night. Too many facts were streaming through his mind plus there was something he couldn't put his finger on. He decided to go for a walk with Ace. Sometimes walks late at night helped to clear his mind and gave him some perspective for solving a problem. Ace was always more than eager to go for a walk, whatever the reason or time of day. Unfortunately, tonight's walk did nothing to answer any of the questions Jack had.

Chapter 13:

When Jack arrived at work the next day, there was a note on his desk from Amy saying that Dr. O'Leary wanted to see the two of them as soon as possible. Jack found Amy at her desk.

"Do you know what he wants?" Jack asked as he showed the note to Amy.

"No, but he did just get back from that DARPA meeting so maybe he wants to update us on what went on there. Let's go see him now, he's been in since seven."

Jack walked right into the professor's office with Amy right behind. The professor looked tired today, his hair was a mess and his tie had been loosened from his neck already. He had a full cup of coffee on his desk but Jack could see it had already gone cold. The professor had his back to the door while he was working on the computer, but he quickly spun his seat around.

"Can you get the door Amy? Thanks. Okay, have a seat you two. As you know, I just flew back from D.C. where I was meeting with the DARPA funding agency. They have agreed to renew our funding for another three years, but there was some bad news.

"First, they are not happy with the progress on the projects to date. They had expected considerably more work out of us by now. You know these guys though are mostly army suits. They don't understand science. I tried to stand up for you two, but they still want some results soon. They are expecting a detailed progress report on your results by the end of next week."

"John," Jack interrupted, "You've got to tell them that we can't be expected to get more research done and if we're spending our time writing progress reports. This is ridiculous."

"I know, but they are giving us a lot of money for this project. And they are paying your salary. Anyway, I need something written from you two by early next week and then I can put something cohesive together. The second problem is that they want us to alter our course of research."

Jack didn't like the word 'alter'. He felt his research project was his domain, and the last thing he wanted was to have some bureaucratic suit in Washington looking over his shoulder, telling him what to do next. He understood this project better than anyone, even the professor here.

"They were excited, Jack, with your idea of using a transporter protein to shuttle the anti-viral protein into cells. They want us to apply that to the anti-toxin project that Amy is working on. The two of you need to work together on this.

"And Amy, I am very concerned with your lack of progress around here."

"What?" Amy looked like she had just been shot in the stomach with a cannon ball.

"Look Amy, you've been here for two years now and you still haven't found an effective protein to block the ricin toxin. I expected that to be done by now. DARPA expected it to be done by now. Now I don't know if you have other problems in your life that are taking your focus from the lab, but you need to concentrate more on your work."

"I can't believe you're saying this, Professor. I am working hard on…"

"Amy." Dr. O'Leary stood up and raised his voice. "I didn't bring you here to Southern California so you could enjoy the sun. You have a Ph.D. in biochemistry. I brought you here because your thesis project and recommendations gave me the impression you could handle a project like this with the pressures associated with it. Now, you have to find that anti-toxin protein. Or," the professor punctuated that last remark with a slam of his fist on the desk before sitting back down. "Or, I'll have to let you go and put Jack on both projects. He's been here less time than you and already he has found the anti-viral protein."

This stunned Jack. It was not like John to lose his temper. And he had never seen him yell at a post-doc in the lab. In science, work and results were not a direct relationship. Jack knew a person could spend twice as many hours as Amy without finding that anti-toxin protein. There was some luck involved. Nobel prize winners weren't necessarily the brightest minds in the field; they were merely bright minds that had benefited from a little luck early in their careers.

Jack looked over at Amy and saw her clenched fists and reddening face. He wasn't sure if she was about to cry or shout. As a friend, Jack didn't want to see her do either right here in front of her boss. Jack rose and pulled Amy up by her arm turning her to the office door.

"John, thanks for the pep talk. We'll get right on that." Jack said.

Jack led Amy out the door, down the hall and outside to the patio.

Amy couldn't control herself anymore and proceeded to let out a tirade on Jack. "Who does he think he is? I am not working hard enough? What about 60 hours a week in the lab is not working hard enough? I spend all my time here working for him. I put all my energy into this project and then he threatens to take it away from me. I should have told him to bugger off, is what I should have done. Or fuck off, as you Americans say. Jack, I should just quit. Maybe I should go home and find a nice safe job in industry."

Amy's temper started to wane and tears were forming in her eyes. Jack grabbed her hand and led her farther from the building. He knew she wouldn't want others to see her crying like this. She was very proud of being a woman in science, a field that was dominated by men at all the top levels.

"I mean really, Jack. Does he think that is a way to get more results from his staff? Maybe if I wasn't teaching his class for him or training his graduate students. Maybe then I could get more work done. He never thanks me for any of that. Some days I'm so tired from all of it, I can barely get home before falling asleep. There's no way I can work harder on this."

"I know Amy. He's not a good boss. But you know how the system works especially here in the U.S. Promotions to professor are made based on research skills, not managerial or training skills."

"Research skills! Ha! He hasn't worked at a bench in years. He would have no idea how to do my work now. It's easier for him to stay in that office and hand down orders. It's our ideas and work that keep him in that position. He certainly doesn't pay me enough to take his abuse."

Jack and Amy sat down in silence for a while. Amy was right about Dr. O'Leary not knowing how to work at the bench. It was one thing to keep current on research and methods, but to implement new methods at the bench could take years to master. Science was becoming more and more specialized with whole teams working on a project so that each individual member could complete their specialty.

"Maybe you should take the weekend off, Amy. I know when I get really stuck, taking time off seems almost counterintuitive, but if you can clear your mind a bit, the solution might become more focused. We could go to the beach and try boogie boarding."

"Maybe you're right. Besides, it would be nice to spend the day with you." Amy brightened up a bit.

She pulled a corner of her shirt up and dabbed at her eyes. "Can we go on your motorcycle? I've always wanted to ride on a motorcycle."

Jack laughed, "Sure. How about I pick you up at noon tomorrow. Bring your suit and we'll rent some boards at the beach."

Jack and Amy walked back to the lab by the back way so they could avoid Dr. O'Leary's office. "By the way Amy, have you looked into that protein library that Abbott labs has?"

"Oh, yes! Thanks for the idea the other day. I called the senior scientist at the protein laboratory at Abbott and he agreed to let me work with the library. But better than that, he is going to do the work for me. He was intrigued by my project and he said he had the facilities and protocols to do a preliminary test on all the protein samples in a matter of a week. Then he can send me the top 10 candidate proteins that appear to work. All I have to do is send him some of the purified toxin, so I am going to do that today."

"See, maybe you already have a break in the project. Things will turn around. And tomorrow you can take out your frustrations on the ocean waves."

Jack returned to his desk to read some papers he had photocopied. The computers were all still in use so he would have to wait to look up Intech. The pressures of the past few days were starting to catch up to him now. Maybe going to the beach tomorrow would really help him too. He would have loved to have taken the weekend off and headed out to his cabin to work on the car, but that wasn't going to happen now. Some days he really wondered what kept him in this field. The pay was low, the recognition was low and the hours and stress were high. It was a wonder there were so many scientists in the world. He had often told friends that this was not a field you chose for money, but rather because you loved the research so much. Actually what kept him going was knowing that his work was helping people. He may not be a medical doctor, but he knew that most of their techniques and tools came from people like Jack working behind the scenes in the labs. Jack was indirectly saving lives with his work. That could make up for an awful lot of money.

Finally Jack noticed that one of the computers was free so he sat down to look up Intech. He started with a simple Yahoo search for the company website. He tried several different spellings but didn't come up with any biotech companies with a name like that. He did find one site that was apparently catering to adult videos. It was amazing what people used for company names these days.

Next, he tried to search the online database of published journal articles. Although the database wasn't complete, it did cover most articles published from about 1970 up to the present. Pretty much every article was listed in this searchable database. Jack searched for the keyword Intech in several spellings but did not come up with any articles. This wasn't too surprising because although most biotech companies do publish good results in reputable journals, newer companies kept much more in house to avoid giving valuable information to competitors. Patents on new drugs or products could take years to complete and clinical trials were often started before a patent was issued.

Jack was discouraged that it was this hard to find out information on a company that was located right here in town. He decided to check one last database of patent applications. Intech would surely have applied for a patent for the influenza drug that they were trying to get approved. Jack searched under the company name but once again came up empty. He even tried several key words looking for applications that would have to do with anti-viral influenza drugs but had no success. This was proving much more difficult than he had anticipated. Brandon must have had the name of the company wrong. According to his story last night, it would not have surprised Jack if Brandon were not being told everything. Maybe Samantha was having better luck searching the city permit databases.

Jack felt he had done everything he could from this end. At least for now. He spent the rest of the day and well into the night working on his anti-viral project. It was after 11:00 by the time he arrived home. He greeted Ace and took him for a walk to clear his mind for the night. Jack realized he was spending way too many Friday nights working at the lab. Amy was right, Dr. O'Leary had no right to come down on her for lack of results. The two of them put in more hours than one should be expected to work at a job.

Chapter 14:

The next day, Jack picked up Amy on his bike. She was wearing blue jeans and sandals with a short sleeved shirt that looked like it had come right off the men's rack at The Gap. Jack could see the outline of a red and white striped bikini under her clothes. She was carrying a bag containing a towel, a Frisbee and what looked like several journal articles. Jack turned off the motor as she approached.

"What are you doing Amy?"

"What? We are going to the beach, right?"

"Right. So why do have these." Jack pulled out the stack of papers from her bag.

"Ohl, I figured I could catch up a little on my reading on the beach."

"Amy, we're going to the beach to have fun, and to relax and forget about work for the day. I command you to leave them."

Amy giggled. "You command me? Come on Jack, you can't make me leave them."

Jack took the papers and walked back up her walkway to her front door and shoved the papers in through her mail slot.

"Okay," Jack clapped his hands together, "Let's go."

Amy put on a fake little pout but quickly grabbed the helmet from Jack. The ride from Amy's apartment to the beach should have been a short direct drive, but Jack took the long way so he could ride along a windy road through the canyons and give Amy a nice ride on the motorcycle. When he finally arrived at the beach, he was lucky to find a parking spot right next to the boardwalk and near the boogie board rental area.

"That was fun," Amy said, "I almost can't wait until we go home."

Jack rented two boards while Amy walked down to the beach to set up her towel and strip down to her swimsuit. The bikini she wore showed off her ample chest and long legs. It looked like a good suit for getting a tan, but not a good suit for boogie boarding in the waves. Well, Jack thought, this could be interesting.

"Here we go." Jack dropped the boards and flippers beside Amy.

"You're not serious, Jack! I didn't think...I mean...you're not serious!"

"You need to get rid of some of that anger, Amy. You can't just bake it away here on the beach. You need to get out there and take out some aggression on the waves."

"But, Jack. I didn't really dress for this. I mean..."

"C'mon," Jack grabbed Amy's board and fins and pulled her up with his other hand to drag her toward the beach. "We'll just take it easy on the small waves. Don't worry," Jack said as he continued to lead Amy into the water.

The small waves were lapping at their ankles as they walked deeper into the cool ocean water. "Now then, you want to walk out into the waves until you can barely touch the bottom. Then slip the flippers onto your feet. We'll paddle out a few feet farther and then we wait for a small wave to come. Use the flippers and your arms to paddle hard to stay in front of the wave. And then just ride the force of the wave into shore. Steer by leaning from side to side. I promise, a few of these small ones, and you'll forget about work, about stress, and," Jack smiled, "about your little swim suit."

Jack was right; Amy really did enjoy riding the waves in on the boogie board. There was something exhilarating about being pushed so hard along the surface of the water without having to do the work. It was a little like riding on the motorcycle, the wind in your face and the sheer thrill of moving so fast. Amy enjoyed it so much that she eventually paddled out to catch some of the larger waves while Jack returned to the beach to watch.

As enjoyable as it was to watch Amy having so much fun doing something new and forgetting her problems,

Jack couldn't help but wish he were here with Samantha. Jack hoped that she had had better luck finding information on Intech than he had. It was odd that with his searches, he hadn't found any information on the company; then again, a company working on a novel anti-viral drug may not want to publish too much on it for fear of giving valuable secrets to the competition. If Jack knew one or two of the lead investigators, he could search using their names for articles they may have published prior to moving to Intech. He would have to remember to ask Brandon for some of that information.

In the mean time, it was good to get out of the lab for the day. Jack knew he worked too hard there, and for that matter so did Amy. They both needed some time away to clear their minds of the immediate problems they were facing with their projects. Jack had found that often a short vacation, even following a series of failures in the lab, could spark some project to work. Tomorrow the two of them would get back to work, and hopefully something would break soon for Amy. He didn't understand what could have incurred the wrath of Dr. O'Leary the previous day. He had never seen the professor so adamant about results. Jack felt he was trying to save the world with his research, but laboratory research wasn't like surgery. He didn't have to save lives in the next ten minutes or even ten days. Expectations were so low with some of these projects, that any success was heralded as breakthrough. Many investigators could spend years, even decades working on a project before the results could be translated to the clinic. Unfortunately, those people typically had a tenured position at a university somewhere. As post-doctoral researchers, Jack and Amy still had to prove to the world that they were worthy of such a position. That was why they worked so hard now.

Jack looked out and saw Amy riding a wave that must have been at least three feet high coming into shore. She was easy to spot wearing that red and white bikini. There were few women boogie board riders out today, but all of them were wearing suits that would stay on during the onslaught of the waves. The wave crashed down near shore and Jack saw Amy stand up and then quickly duck back down. Yep, she had lost her bikini top in that wave. She quickly replaced her swimsuit and walked back up to Jack on the sand dragging the boogie board behind here like a hunter returning from a successful kill.

"I saw that."

"Oh, yeah, that happened a few times. I guess this isn't exactly the most practical suit for this. I can't believe I never tried this before though," Amy was beaming with a smile. "That was so much fun. Of course we can't do this in England, cold water and all."

"Well, here's some fresh water to drink. You might want to get the salt water taste out of your mouth."

Amy lay on the towel next to Jack and the two of them relaxed under the watchful stare of the sun with the ocean breeze blowing across them cooling the air.

Chapter 15:

An hour later, Jack woke up and nudged Amy awake. "Hey, you want to get some lunch?"

They dressed and headed back to the motorcycle, packing the beach gear into the saddle bags before returning the boogie boards. Without a destination in mind, they left the beach, driving down the coast toward the downtown area. As they passed through the downtown, Jack remembered a place that was a few miles south of the city. Amy didn't seem to mind the ride. She was looking around enjoying the sites. They passed through the downtown on the freeway and then headed southeast out of the city and into the more deserted neighborhoods.

Jack pulled into the dirt lot of a bar called Ridin' Inn. The lot had only two cars in it with the rest of the space occupied by motorcycles, mostly Harley Davidsons. The sign was covered in a thick layer of dust that had been kicked up over the years by passing motorists. The windows along the lower floor of the building were covered with vertical rust-stained bars that almost matched the color of the solid wood door just off the parking lot.

"Jack, is this a biker bar? I've always wanted to go to one of these."

"Umm, yeah. I guess you could call it…"

"Ohh, look at this one. The tail pipe is shaped like a snake." Amy pointed to a very original Harley Davidson painted bright red with green streaks coming off the gas tank. The exhaust pipes coming out of the engine had been pounded and bent to resemble the shape of slithering snakes. The ends of the exhaust had replica cobra heads with silver steel tongues licking out toward the rear wheel.

"That's Snake's bike. Come on, I'll introduce you."

Jack and Amy walked through the heavy oak door of a building that was probably at one time a house. The front room was large and dark. There was neither sunlight nor breeze coming into the bar, and it didn't appear that either had seen the inside of this place since the days when people had lived here. The bar took up the entire far wall and most of the people were lined up along it drinking. To the left was one pool table with a low hung light advertising Bud Light to illuminate the players. Several round tables, most of them empty, were scattered around the room. Jack felt an arm reach across his throat from behind. Amy screamed and Jack felt a poke from something the size of a gun barrel into his kidneys.

A voice made rough and low from years of smoking spoke from behind him, "You don't belong here mister."

Jack looked over at Amy and saw terror in her eyes. He simply smiled and made no move. "Snake, get your dirty hands off me. You smell like a gas station."

Snake spun Jack around and gave him a bear hug. "Good to see ya too, Jack. Wha's been keepin' you from comin' around."

"Snake, you're drunk already, aren't you?"

"Ain't not. But good idea, le's get another round. Hey Patty," Snake yelled to the bartender, "get me two, no three more for my friends here."

"Snake, this is my friend Amy. And keep your hands off her."

"Sorry man didn't know you had a lady here. Nice to meet you, Amy."

"She's not my lady, but she is a very good friend."

Jack left Snake at the door and headed up to the bar with Amy sticking right next to him. Amy reached out and gripped Jack's arm in a vice-like grasp.

"Trust me, you don't have anything to worry about. I know most of the guys here, and they wouldn't do anything. Hey Patty, can you put on a couple of burgers and some fries for us?"

"I think this is about as far from the Hoffman Institute as I could possibly get. If you are trying to get me to forget about work, you've done your job, Jack."

Snake came up to the bar and sat down next to Jack. "So, didja get that seat you wanted?"

"Snake, I know you were just outside checking out my bike, so you know I didn't. I'm not sure yet if I want to. I've broken this one in. I don't want a stiff new seat under my ass."

"Well, if you change yer mind, just ask."

Jack looked around the room, recognizing a few other people at the bar as regulars at the bike shops. Tattoos were about as plentiful as the beer here.

"Snake, I may have a problem, and I was wondering if you could help me."

"Jus' say the word, brother."

Jack smiled and turned to Amy. "Would you mind if I talked to Snake at a table for a bit?"

Amy looked around nervously. "Don't worry lil girl. Patty'll watch ya." Snake told her.

Jack and Snake walked to the side of the room and pulled up to a table. "Snake, I have a friend that may have gotten himself into a bit of trouble. I am trying to help him out a bit, but I think I may have gotten myself into something too." Jack took a sip of his beer and set it back down on the table. "Now I don't know what trouble or with whom, but I need a little help here. I think someone has been following me. Several times in the past few days, I've noticed a silver Nissan at work and then occasionally I see it following me. I saw the car while on my way home from work, during a ride up the coast and even at the beach today. Is there anything you can do, maybe find out who this is?"

Snake just sat in his chair without moving. It was so dark in there that Jack wasn't sure if Snake was still awake until he lifted the bottle to his lips and took a drink. "You know I'd do anything for ya." It was amazing that Snake could go from stumbling drunk to coherently sober in minutes. "I'll get a couple of the guys together and we'll see what we can do. What do you want us to do to this guy?"

"Nothing yet, just, when you see him following me, watch where he goes. I think I may know, but I need you to follow him a bit. Thanks Snake, I owe ya."

"No, I don't think so. I think we all owe you far more than we could do for you. You took a real chance that day coming to court to testify for us. You didn't owe us anything and you aren't even with us. You stood up that day. Anytime you want in, just let me know. We owe you. Now, tell me about this girl you got with you. She's a looker I tell ya."

"Uhh, I guess she is. But she's just a co-worker. We needed a day away from work so we headed out to the beach for a bit."

Snake whistled. "To the beach? Man you got it for her huh? And she's been watching you ever since you came over here."

"I don't have anything for her. And she's watching this way because she feels uncomfortable here. No, I got this neighbor I've been hanging out with a lot lately."

"Yeah, well, you can never have too many."

Jack punched Snake in the arm and stood up, "You're sick Snake. I'm gonna eat, but thanks for everything."

Jack sat back down at the bar just as Patty was setting the burgers down. Amy dived into her burger looking as though she hadn't eaten in days. That was what swimming against the waves would do to you. "So how did you find this jewel of a place, Jack?"

Jack smiled at Amy. "Long story."

"I've got time. How do you know these people."

"Okay, but I don't need this story getting out okay?" Amy nodded in between bites. "Snake owns a custom bike shop a few miles east of here, and I went out to his place a few times to have him fabricate some parts for my bike. It's really hard finding parts for such an old bike. We got along pretty well and used to go for rides on some of the country roads in the mountains east of here. One time we were out on a ride and a member of a rival gang was gunned down. Apparently

Snake and some of his guys had threatened to kill this guy in a pretty public place, and there were planty of witnesses. The cops arrested Snake and three other guys and charged them with first-degree murder. You've got to understand, these aren't the most law-abiding guys around and they probably have committed many crimes. The cops wanted to hang Snake for this crime. Snake knows that anyone associated with him is a target by the police so he never would have asked me for help, and I didn't owe him anything because I barely knew him. Anyway, I went to court and testified that I was with them riding on country roads all day when this guy was killed, so Snake couldn't have done it. The police tried to poke holes in my story, and they investigated me for about three months before they gave up. Snake feels like he owes me for that one.

"We both know we're from different worlds so he would never ask me to pledge with his guys, but we get along now because of that. These guys may not be the finest citizens, but they do stand up for their friends. And they're the only guys I know that love Harleys as much as I do."

"So what did you ask him about today?" Amy asked.

"For that story, you'll have to join me for dinner later." Jack smiled. On second thought, if Snake was right, and Amy really was interested in him as more than a friend, maybe he shouldn't have asked her to dinner. Well, he had agreed to take her mind off work for the whole day.

Walking back out into the afternoon light was momentarily painful. As the all-white scene faded back into colors and shadows, Jack saw that the sun was sinking lower in the sky. He dropped Amy off at her apartment and headed back to his place to shower and change for dinner.

Chapter 16:

As Jack was heading back out of his apartment toward his Jeep, Samantha was walking up the sidewalk. He had wanted to see her all day, but right now, he felt he didn't want to talk to her. He had enjoyed his day off. It was more than a day away from work, it was a day away from the stress and worry of the past week. Samantha was a reminder of that stress.

"Jack, I've been trying to get a hold of you all day. Brandon is still in Seattle, but I've got some news about Intech."

Jack put his keys back in his coat pocket and walked with Samantha to her apartment.

"I looked up Intech in the city records like you suggested. They filed a building permit for the building in the warehouse district that you described. They list their corporate headquarters in Santa Rosa. I just don't know what this will tell us though."

Jack thought about that for a minute. He had hoped that between the two of them, they would find some information on Intech and their new influenza drug. There was still something that was tickling the back of his brain.

"…out anything?"

"What? Sorry Samantha, I was just thinking about what you said."

"I said, did you find out anything about Intech?"

"No, sorry. Nothing came up on them with my search"

"Well, wasn't that a bit odd? I mean you said you would find articles on their work on the anti-viral drug"

"I had hoped to find articles published by them, but there could be many reasons why I didn't. The company may be very new with most of the work published before the company name would have been attached to the papers. Second..."

"Actually, the building permit for Intech was filed just two years ago. So you may be right on that part."

"Second, they may be delaying publishing their results until they can get a patent issued."

"I thought scientists were supposed to publish everything they worked on so that others could learn from it."

"Generally that's true, but lately more publications get delayed until patents are issued or until investigators can get a lead on the competition. So it wasn't that unusual not to find anything on them.

"But, I've been thinking, no one in their right mind would think that the FDA would approve a new drug following the release of a virulent influenza over a whole city. There are other types of influenza that don't cause serious disease in humans but are still very similar to the virus that does cause disease. My guess is that they are planning to release one of these strains, and all our worry would be for nothing."

"How can we find out?"

"I was thinking if Brandon can get me into the building, I might be able to find out some more information. Maybe I could even get a sample to examine back at my lab. Better yet, if I could get a sample of their anti-viral drug, I might have a better understanding of what they are planning to do. We'll just have to wait until Brandon gets back to see."

"He'll be back late tomorrow night so maybe we can get together with him Monday night. Oh, I'm sorry." Samantha had just noticed the nice pants and jacket that Jack was wearing and realized he was heading out when she caught him.

"Oh, yeah. I'm supposed to go out with a friend tonight. But I can call and cancel if you want some company."

"No, no you go Jack. I'm sorry I kept you."

Jack didn't want Samantha to think he had another date. "She's a co-worker named Amy. We've had a rough week at work, so we wanted to go out for a nice dinner. It's nothing really, I could cancel."

"Jack, go. I'll be all right. I need to write some anyway. I've been getting way behind worrying about Brandon all the time."

Jack kissed Samantha lightly on the cheek and went out to his Jeep.

Chapter 17:

Amy was waiting on her front step as he drove up. Jack had only seen Amy in the lab where most people wore jeans and a t-shirt most everyday. Tonight she was wearing nice slacks and with a long black jacket over a white silk shirt. She looked very disappointed as Jack got out of the Jeep. He hoped she hadn't been waiting too long.

"Jack, I thought you'd bring the motorcycle again."

"I thought you might get too cold riding at night. Next time, I promise."

"So where are we headed?" Amy asked as she got into the Jeep.

"I thought we could head to Little Italy to a nice restaurant I found there. You do like Italian?"

"Of course. And you're going to tell me what you talked to Snake about?"

"I promised."

Little Italy was a small neighborhood in the city that had once been populated by Italian fisherman when this big city had been a quaint fishing village. The neighborhood was quite distinct from the rest of the city. Many of the houses and buildings that survived to this day had strong Italian influences in their construction. Little Italy had become fashionable lately with a lot of new construction going on, but there was a general consensus to retain the Italian influenced architecture. The small street with large sidewalks reminded many people of Rome with delicious aromas of olive oil and garlic wafting across the streets from the numerous restaurants. Jack knew of one behind the main street that sat only about 100 people but had probably the best Italian food he had ever experienced.

"So Amy, where are the good English restaurants in the town?"

"A good English restaurant is an oxymoron. Even the English can't stand English cuisine. We eat mostly Indian food nowadays with some French and Italian too. I mean come on, would kidney pies and warm beer appeal to you?"

The maître d' seated them in the back corner of the main room. This was off the tourist season so the restaurant was only about half full tonight. Jack ordered a bottle of Chianti and an appetizer of mozzarella, tomatoes and basil before beginning to tell Amy what his past week had been like.

"Before I tell you what I told Snake, I need to start a bit earlier. I need you to keep this to yourself though because some people could get hurt if it gets out that I know or that I told you. Okay?"

"Jack, you're scaring me here. I though you were just talking motorcycles."

"Amy, I need to know you will keep this to yourself. Otherwise we can just have a nice dinner and talk about something else."

"Okay, I can keep a secret. You can tell me anything. That's what friends are for."

Jack paused as the waiter delivered the appetizer and the bottle of wine. Jack took his glass and toasted the evening, "To letting work problems go for a day."

After a sip of the wine, Jack relayed the events of the past week to Amy stopping short of explaining his interest in Samantha. For some reason he couldn't bring himself to tell Amy, his best friend, that he was falling for Samantha.

"The other night we confronted Brandon about this. At first he tried to deny he was involved in anything, but eventually he told us everything he knew. He told me the building I had followed him to in the warehouse district was owned by Intech." Jack saw what looked like recognition in Amy's eyes. "Have you heard of this company?"

"No," Amy said, "I don't believe that I have."

Amy seemed to digest all the information that Jack had just told her. She was no longer interested in the food that was cooling on her plate and instead was looking at him with a concerned expression on her face. "Do you think they're planning to infect the whole city by releasing the virus into the wind?"

"Well, I thought so at first, but now I'm thinking that would probably pop up too many red flags with the CDC; a mass epidemic of the flu, all with the same starting point. My guess is that they are going to release influenza B, which as you know only causes very mild symptoms at its worst, unlike the far more common and serious influenza A."

"Do you think Brandon will go to the police?"

"No, he's gotten himself into some deep shit. Sorry, excuse the expression. Anyway, stealing money from these guys was a bad idea. They've threatened him and his family if he tells anyone. That's why you need to keep this to yourself. I've been trying to help any way I can, but now I think I need to get inside that building to find out what they have and what their plans are."

Jack and Amy ate their dinner in silence while Amy thought about what Jack had just told her. "Is there anything I can help you with?"

"Thanks, Amy. I don't know yet. I told you all this because I think that I might need your help soon. If I get a sample of the virus or the anti-viral drug, I will definitely need your help in identifying what I have. You are one of the best protein biologists I know and besides, I trust your judgment in the lab."

"Jack, take care of yourself." Amy reached across the table and took Jack's hand in hers, "You're a scientist, not a detective. You Americans all think you are cowboys. You don't have to save the planet by yourself. You're probably right and this is nothing. Besides, if Brandon got mixed up with these guys and even stole money from them, how do you know he is telling you the truth on this."

Jack thought about that last part a minute, "I talked to him myself. I just don't think he was lying about this. Amy, I will be careful. And I know I'm not a cowboy. I don't even have a trusty steed." Jack said as he laughed.

The two of them finished their dinner and then Jack took Amy back to her apartment. "So I'll see you tomorrow at the lab?" Jack asked.

"Yes, I will be there as usual. Jack," Amy paused, "do you want to come in for a drink?"

"Aww, thanks Amy, but I really should get back home to let Ace out. Besides I'm about done for the day. Maybe another time."

Amy looked disappointed, "Sure, another time. Thanks for a great day Jack. You were right, I did need some time away from work to relax. I would love to do this with you again sometime."

During the drive home, Jack tried to think of a way to get into Intech. He hoped Brandon would have an idea for getting him in. He would probably need an hour or more to give him time to look around and maybe find the freezers where samples would be kept. Night would probably be the best because fewer people would be around to question him. He would try to get together with Brandon as soon as he could. In the mean time, he had plenty of work to do in the lab.

Chapter 18:

The next three days seemed like a blur to Jack. He worked long hours in the lab with Amy right at his side. While she was waiting for Abbott labs to get back to her on the anti-toxin protein search, she had time to help Jack with his anti-viral research. His work was coming along great, but he tried hard to avoid Dr. O'Leary. The last thing he wanted to do now was give a good progress report on his work while Amy still had nothing going on her project.

It wasn't until Wednesday night that Jack saw Samantha again. He arrived back home at 6:00 after having taken Ace for a walk when he saw Samantha stepping out of her apartment.

"Hi Jack. You must have been working hard lately. I haven't seen too much of you. I was just coming over to see how you were."

"I was just about to make some dinner. Would you like to join me?"

"I just wanted to say hi. I don't want you to go to any trouble."

Jack smiled, "I promise I won't go to any trouble on your account tonight. I was going to make chicken Parmesan and I have three pieces of chicken. If you want, you can bring something over and we can just split dinner chores tonight. I did want to talk to you, too."

Jack took out the chicken he had thawed and rubbed it in melted butter and then dipped it in a mixture of Parmesan cheese, dill and oregano. He put the chicken in the oven to bake while he opened a bottle of Merlot. As he was setting the table, Samantha knocked on the door and brought in a salad and a bucket of ice cream.

"The chicken will be ready in about 15 minutes, so you are right on time. How about a glass of wine?"

"I just want to say I'm sorry for falling apart on you last week. You were a good friend to put up with me like that. I talked to Brandon on Monday and he sounded good. His trip to Seattle went fine and I think they are asking him to do less work for them now. He says they still want him to take wind measurements, which is fine by him since he likes to surf. I told him you wanted to talk to him, but I didn't mention the reason."

"When do you think we can get together with him?"

"I was supposed to see him later tonight, so you could come over then."

Jack served the chicken and the salad as the conversation drifted to more date-like topics. Samantha explained that her book was coming along pretty well. She felt she had a good handle on the plot and was writing an average of five pages a day. Jack told her about his Saturday with Amy and their recent progress at the lab.

"No wonder I haven't seen you lately, Jack. You've been pretty busy. I can't believe you accomplish some of the things you do in the lab. I don't think I've ever known anyone so smart before."

Jack blushed, "It's really not as hard as most people believe. I was just trained in this for so many years that I can do it now. I could never work at the paper or write a book, because I don't have the education for that work."

"Yeah, but I think I would never have been able to make it through the college courses you took, and I know I couldn't have made it through graduate school. I would never have stuck with it that long."

"You would have if it had interested you. I wouldn't have made it through one year of all those English courses. That stuff really bores me. I mean…" Jack blushed and looked up at Samantha.

"Touché, Jack."

Jack spent the rest of the evening trying not to put his foot in his mouth again. After ice cream, they headed next door to wait for Brandon.

Brandon knocked on the door, and then walked right in without waiting for an answer from Samantha. "Oh, sorry guys." Brandon looked unsure of what he should do. "I can leave you two alone."

"It's okay Brandon," Jack spoke. "We just had dinner and were waiting for you."

"Sam told me you wanted to talk about Intech." Brandon entered the apartment and took a seat in the chair where he had sat last week when Jack questioned him on his role with Intech.

"I've been thinking about what you told the two of us last week regarding Intech's plans for the 'rollout' of their new anti-viral drug. I think we can all agree that releasing a virus on an unsuspecting population sounds vaguely like something the government would have done back in the fifties. But I'm not sure anyone would believe they could do that today. That's why I'm not sure that their plans include infecting the whole population and may not even include the use of a virulent influenza virus."

"What do you mean virulent virus?" Brandon interrupted.

"A virus that actually causes disease. There are three different types of the flu virus and only one actually causes severe disease and can kill people. The other strains of influenza can infect humans, but don't really cause any disease."

"So how do we find out what they are going to do?"

"That is why I wanted to talk to you. I was thinking that if I could get into Intech, say late at night, I might be able to get a sample of their virus. I could take this back to my lab and determine what kind of influenza they are planning to use."

"I'm not sure how to help you with that. They have security guards there at all hours, and there are a lot of cameras around. They're very paranoid about security. They make all the people that work there park blocks away and walk in."

Jack sighed. He was afraid of this and hadn't thought of any way to get around this security. He had hoped that Brandon would have an idea.

"Wait, I might have an idea," Brandon said, "There is a late shift of scientists that works from 9:00 PM to 9:00 AM. I've been there when the shifts have changed. I think the later shift is mostly technicians that monitor the machines and keep things going, because they don't look as nerdy as the scientists that are there during the day. The problem is the only way to get in past the guards is with a badge."

"Well, don't you have one?"

"No, I only go there with Aded, and he has a badge. I have to go there tomorrow, and I can probably get a badge then. Most of the time the badges are just pinned to the lab coats. I could probably grab one from a coat that is lying around. Then, if I were you, I would go on Friday night. I remember from talking to people there that the security guard staff is smaller on the weekends when there are fewer people in the building."

"Do employees still come in on Friday night?"

"Yeah, but not as many. They have a rotating work schedule so they work every other weekend. It might make it easier for you to get around then too. There will be fewer people around to see you."

"Are you sure you want to do this Jack?" Samantha said with obvious concern in her voice.

Jack thought about it for a minute. From the scene he saw in Santa Rosa with Brandon, he really didn't want to get on the bad side of Saresh. He knew he was putting himself in some danger, but he really believed that it wasn't too big a risk. Besides, he needed to get a sample of this virus to find out what Intech was planning. If they were going to release a large dose of virulent influenza over the city, he would have to find a way to let someone know even if it meant a risk to Brandon and Samantha. If they were only planning a small release of a less virulent influenza, maybe he could keep his mouth shut and protect Samantha's family. What it really boiled down to was determining Intech's plans without risking Brandon's or Samantha's life.

Jack looked Samantha in the eyes and reached for her hand, "I need to do this Samantha. The more we know about Intech, the better prepared we can be for their plans." Jack turned to Brandon, "I think this is as good a plan as we are going to get. Get me that badge and I'll try to get into Intech Friday night. But be careful."

Chapter 19:

The next evening Samantha dropped the ID badge through Jack's mail slot so when he came home he had a whole evening to work on changing the face on the badge to his own. The badge was the size of a credit card in length and width but was about three cards thick. On the front it simply said Intech with a small logo of a virus with an anti-body bound to it. Below this was a picture of a white man with short brown hair. Perfect, Jack thought, the resemblance was already pretty good to his own face. The thickness of the card though implied some sort of magnetic code that would probably allow the owner to enter certain facilities and not others. Hopefully, the owner of this card had access to at least the freezers.

Jack noticed that the picture in the photo was taken in front of a dark blue background. He looked through his sheets and found one that was pretty close in color. He then set up the digital camera on a tripod so that he could take his own picture. On the computer, he was able to adjust the size of the photo to match the one on the card. Then he printed the picture on his best quality paper. With a hobby knife, Jack peeled the top layer of plastic off the card and removed the photo from underneath. He then placed his photo on the card and used clear packing tape to re-cover the card. The card looked just like it had originally but with Jack's face smiling back at him.

Friday at work, Jack was anxious most of the day. He began to rethink what he was getting himself into. He really had no idea how to sneak into a place or how to steal a virus sample once he was there. He had hoped that he would figure it out as he got to that point, but now he realized he had better have a plan. The virus and anti-viral drug would probably be frozen at $-80°$ Celsius to prevent degradation. He would have to keep the viruses at a temperature of at least $-20°C$, which was a normal freezer temperature. Dr. O'Leary's laboratory received several shipments each day that arrived in a cooler packed with dry ice. That would be sufficient for transporting the virus, but he couldn't walk past the guards going in or out with a cooler tucked under his arm without arousing suspicion. He could use a small cell freezing apparatus from the lab. They were the size of an orange and could easily be slipped into his coat pocket. If he froze one of these small coolers to $-135°C$ and then kept it on dry ice until he got to Intech, he could place the virus and anti-viral samples inside it until he got back out to his Jeep. That might work as long as he didn't have to spend too much time inside the building.

"Hey Jack," Amy called from across the lab, "I just got off the phone with Abbott labs. They found four hits in their protein database that block the activity of the ricin toxin. They are going to send the samples to me so that I can work on them."

"That's great news! See, I told you something would break on your project soon."

"Yes, but this was your break. You told me to use Abbott labs for their protein library. I owe you big for this. How about if I take you out to dinner tonight."

"I can't, sorry." Jack motioned for Amy to follow him outside the laboratory doors. "Remember last weekend I told you about Intech?"

"Yes, I remember. Something about a friend of yours getting in trouble."

"That's right. Intech is the company that I heard had plans to release the influenza virus on the city and then roll out their new anti-viral drug. Well, I'm going to try to get into the company tonight to get a sample of the virus and the drug so that we can figure out what their plans are. I'll bring it back here to work on. We can start on this tomorrow."

"Jack, I have to say again, do you really think this is a smart thing? I really don't think you should be doing this. You're not a burglar, how do you think you're going to get in without them noticing?"

"I have an idea. Trust me Amy. This is the best way."

Chapter 20:

Jack walked back to the lab to finish up his work for the day. He placed the cell freezing apparatus at −135°C and put some dry ice aside for later. At 9:00 PM he pulled out the cell freezer, put it on the dry ice inside a cooler, and went out to the parking lot. He had driven the Jeep today because he felt it would be less conspicuous than the Harley. As he approached the parking lot, he saw the silver Nissan 300ZX on the street with two men inside. Jack ducked quickly back behind a side building. Walking behind the building he could stay out of view almost all the way to his Jeep. The men were parked near a streetlight, and as Jack came back out into the parking lot, he could see they were still looking in the direction of the laboratory. Maybe they hadn't noticed that he had left yet. He walked with his head down to his Jeep and got in. There was another large motorcycle parked in the lot. Anyone who knew motorcycles would know that it wasn't his, but maybe these guys thought he had ridden his bike in today. Jack drove by the 300ZX without attracting any attention, although he continued to search his rear-view mirror the entire drive downtown and into the warehouse district.

Jack parked in the location that Brandon had said was reserved for the staff and walked the three blocks to Intech with the cell freezer in his lab coat pocket. His heart was pounding now; he had no idea what he would say or do if they caught him. He continuously fingered the pass in his pocket hoping this would be his free ticket into the building. At the dark glass front doors to Intech, Jack paused and took a deep breath. There was no turning back from here.

Jack walked through the glass doors and saw a hallway with another glass door at the far end. As he approached the end of the hallway, he saw a small window on his left with a security guard watching several monitors. Jack pulled out the badge and flashed it at the guard hoping that if he didn't give the guy too much time, he wouldn't notice the modification to the card.

"Hi Doc. Back so soon."

Jack froze. He had hoped that there would be so many scientists here that the guards wouldn't be able to recognize him. The good news was that he looked enough like the owner of his card that he could pass for him. Jack knew that Brandon had gotten this card from someone on the day staff.

"Yeah," Jack looked at the name tag on the guards shirt, "Bernie. I need to monitor my cell culture growth assay all night. Just had time to grab some dinner."

Jack kept his head down the whole time and now bent down to pretend to tie his shoes. He hoped that if he gave the guard a complicated enough explanation he might not want to talk anymore. Jack was right and Bernie buzzed the door open at the end of the hall.

"You guys work too hard I tell ya. Have a good one Doc."

"Thanks, Bernie, you too."

Jack walked quickly through the doors to get out from under the eyes of Bernie. Of course now that he was out of the hallway he wished he had taken a better look at the monitors to see where the cameras were placed. The doors opened into another long hallway with the word Laboratory printed on the door at the end. Jack walked through the hallway into the lab.

Here he finally felt a little relief. This was where he spent most of his waking hours, he felt comfortable in a setting like this. He was looking at the lab from the opposite side of the loading dock door that he had peeked through last week. There were several lab benches lined up approximately 50 feet long and three benches deep. Rising from the top of the benches were shelves containing various bottles and plastic tubs labeled with chemicals and other common lab solutions. Scattered across the benches were experiments in various stages of completion with a power supply on the bench closest to Jack. Around the outside of the room were desk carols set side by side probably for the scientists that worked in this room.

There was only one other person in this room now and she was on the far side with her head buried in one of the flow hoods. He saw several isolation hoods on the other side, but these looked more like chemical hoods than the more rigorous cell culture hoods. He would probably find the cell and virus hoods in an isolated room. Jack turned right and walked past a whole wall of large ultra-centrifuges that resembled clothes dryers. These could spin at 120,000 RPM, fast enough to spin viruses out of a solution. He continued past the benches and was amazed that all the equipment looked so new. Most labs were built up to this size over a period of decades so equipment would be in various states of disrepair and age.

Jack left the lab and entered a hallway with several doors on either side. He walked past a door labeled Microscopy. That was probably the room that housed the expensive microscopes capable of visualizing fluorescent dyes and maybe even an electron microscope that took such detailed pictures that individual viruses could be seen. The next door had a window, and he could see two rows of isolation hoods. This would be a biosafety level 2 room where they could grow cells and most viruses...those for which there existed a treatment or cure. He debated walking into that room but decided against it. He didn't want to take cells or viruses that were kept at 98°F. He would have no way of getting them back to his lab while keeping them alive. He needed to find the viruses that were already frozen. He had to find the freezers.

The next door was labeled Stockroom. Jack looked inside, but it only held boxes of lab supplies. He saw one small freezer and refrigerator that probably held those supplies that needed to be kept cooler. The next door had no label on it, so Jack decided to enter that room.

Bingo! This room had two oversized air conditioners placed right in the middle to prevent the equipment from overheating. All around the outside of the room several large freezers hummed away in an attempt to reach the ultra-cold temperatures that were required to put cells and viruses in a cryogenic state. Jack walked around the room examining the freezers. He saw two that held a temperature of $-80°C$ which was cold enough to store viruses and proteins like an anti-viral drug. The next three freezers were shaped like chest freezers with a digital display reading $-135°C$. These freezers held cells in a cryogenic state until they were thawed and grown again. After these, and back around toward the door, were several large cylindrical containers that held liquid nitrogen. At $-170°C$, these would also hold cells but for a longer time than the $-135°C$ freezers. In the middle of the room between the air conditioners was a large liquid nitrogen tank used to refill the smaller tanks with the cells in them. Jack looked around the room without seeing any cameras, but he didn't know if there had been any in the hallway that would have recorded him entering the room. If he stayed here too long, someone might get suspicious.

Jack looked around for a log-book that might indicate what stocks were kept where so that he could get what he came for and leave. In the meantime, he removed the cell freezer from his pocket; his leg was starting to get very cold. On the shelf above each freezer, Jack noticed a thick three ring binder. He pulled down the one over the first −80°C freezer and thumbed through a few pages. Each page identified the contents of one box. Jack quickly opened the nearest freezer door and looked in. There were slide out shelves throughout the freezer with 6 shelves wide and 18 high. Pulling out one shelf, Jack counted 4 boxes in a shelf. That was 408 boxes per freezer! Jack's heart sank as he thought of the chances of finding the virus samples. He looked back at the book he held in his hand and noticed that every vial was listed by some code. The box he was looking at had vials labeled V2-55 through V2-58 with 8 to 10 vials of each label. Without a key, this could be an impossible task. He guessed that V stood for virus, but that wasn't going to get him too far.

Suddenly, Jack heard voices outside the room. Maybe he had already been in there too long. The voices receded down the hallway, but not before he thought he heard his name. And not the name on his ID badge, his *real* name. He would have to be careful. Jack opened the door a crack and looked out into the hall. When he saw no one there, he slipped out and walked down the hall farther. The next door had no label but Jack could see through the glass window that there were several computers. Maybe he could find the key to the vials on the computer system.

No one was in this room, so Jack picked a computer that was against the wall of the hallway and away from the window of the door. He hoped that he would not be visible if anyone were to look inside the room. Jack searched the hard drive for Project Windswept, without finding anything. That would have been too easy. Next he searched for influenza and anti-viral without finding anything. Finally he searched for virus and found several files with one labeled virus stocks. He opened this file which turned out to be a database of the freezer stocks with labels for each freezer as well as location within the freezer. Unfortunately, each location was identified by the same code he had seen on the vials he had pulled out. He found one box with vials labeled V1-21 — V1-23. Without a key, he would have no way of knowing out what these vials contained.

Jack checked his watch; he had been at the computer for almost a half hour. He quickly looked around the room and saw a camera in the far corner. It would have been able to see the whole room. He realized now, he should have looked for that before he sat down. Did they know he was in here? Just then the door to the computer room opened and a man in a white lab coat stepped in and walked straight toward a computer at the back of the room. He decided that this might be a good time to leave. Obviously there wasn't much information to be learned from the computer files.

Jack returned to the large laboratory. Maybe he could find a lab notebook with some indication of what he was looking for. He really didn't want to be in here any longer than he had to, but it would be fruitless to leave now without getting what he came for. He knew he wouldn't have another chance like this. The laboratory was lined around the outside with desks for the scientists. Each desk was about 3 feet wide with shelves lined with textbooks and vendor catalogs. This looked familiar to Jack. He walked slowly around the outside of the room looking for the desk with the name on his badge. He found it along the far wall, near the loading dock doors.

The laboratory book was sitting open on the desk so Jack sat down and started thumbing through the list of experiments. It could be hard to follow, since this person was obviously doing several different experiments at the same time. He saw some mention of virus production with dates and labels similar to those he had seen on the vials. He noticed though that the viruses detailed in this lab book were all labeled V2 and sequentially increased the second number. Maybe V2 was the second generation of virus with the second number indicating the batch number. Jack looked on the shelf and saw a previous laboratory book. Looking through this book, he found several entries for V1 in the beginning with the first entries for V2 near the end of the book labeled V2-1, V2-2 etc. Now if only he could find out what V1 and V2 were. He decided that the best course might be to get a vial of each. He hadn't seen any entries for the anti-viral drug; this scientist must be working only on the virus.

Jack headed back down to the freezer room and started collecting vials. He took vials labeled V1-25, V1-33, and V1-35. And then the V1 numbers stopped so he moved on to the V2 vials taking batch 2, 21, 55, and 76. The numbers continued up from there. Since there weren't any cameras in here, Jack looked through some of the other freezers. He found in the −135°C chest freezer several towers that had to be pulled up. Each one of these towers held 12 boxes full of vials. Most of these were obvious names of cells that Jack recognized as well as several boxes with the virus stock names on them. They would have put some of the viruses at −135°C to keep them in more long-term cryogenic storage. Jack was beginning to think this trip was a waste of time. What was he thinking? All of his stocks were coded too. The vials were small cylinders about the size of a grape so that label such as 'This Is The Anti-viral Solution' wouldn't fit. Well, he had several vials that he thought were the virus stocks, so maybe it was time to get out before anyone found out who he was.

Jack slipped out of the freezer room. He didn't notice the flashing red light on the big upright −80°C freezer that indicated the temperature had increased while Jack had the door open. Jack didn't know that the freezer was wired to the security guard's booth where they monitored more than just the cameras.

Chapter 21:

After leaving the freezer room, Jack headed for the exit, but then changed his mind. As long as he was here, he might as well check out the rest of the building. He knew he was only seeing about half of one floor. After the freezer room and computer room came another storage closet, but this one was full of supplies used for the BSL-2 cell culture room he had seen earlier.

The next door he saw was set back from the hallway, which is why he hadn't seen it earlier. The window on the door was blacked out so Jack couldn't see in, but he could read the label on the door that read BSL-3 room. Jack remembered his first visit to Intech. When he had been peering in through the loading dock door, he had seen signs indicating this room. There would have been no reason for a BSL-3 facility to grow or study influenza. These rooms were only necessary when working with microbes that could cause death in humans with no known treatment. What were they working on that would require a room like this?

Jack tried to open the door but found it wouldn't budge. He looked around and found a magnetic scanner next to the door. He tried swiping his badge over the scanner and the door opened. He entered an anteroom which had gowns lined up along the left wall. Foot covers, facemasks and hairnets were along the other wall. A sink was right behind the door he had just entered. At the far end was another door that led into the actual laboratory. Next to this door were boxes of gloves labeled S, M, L, and XL. Jack pulled on a gown over his lab coat and slipped on the other protective gear. The next door opened with some difficulty because the negative air pressure in the BSL-3 lab would suck air into it. This prevented any air contaminant from entering the rest of the lab. Jack walked into a large room that must have occupied the rest of this floor. In the middle of the room he saw two rows of isolation hoods with at least eight hoods in each row. Along the outside of the room was a row of incubators stacked two high for growing cells and viruses.

Jack heard a noise inside the room and felt his heart skip a bit. He didn't want to see anyone in the room that might know he didn't belong here. It was a long way out from here. Just then a robot on wheels came around the corner. The robot opened the incubator door nearest him and pulled out a large cylindrical vial that had been lying on its side and was half-full of red liquid. Jack knew this was a method for growing a lot of cells. The incubator was full of these vials which were about 2 feet long and 6 inches in diameter. Each vial lay on its side suspended by rollers that continuously rolled the bottle so the liquid would cover the whole inside at all times. The cells grew along the inside of the bottle, and the liquid was actually growth media for the cells similar to blood. Each vial held at least one billion cells with each cell capable of producing at least 1000 viruses. If this whole room were full of those bottles, the virus production here would be phenomenal.

Jack followed the robot toward the back of the room and watched as it put the bottle upright into another machine. This machine gripped the bottle and spun the top off. A tube was inserted into the bottle, and the liquid was siphoned off and into the machine. New, fresh media was added back into the vial, and the top was spun back on. The robot retrieved the bottle from this machine and moved back to the incubator. This was an extremely high tech automated system for growing lots of virus. Jack had never seen anything like this before. He knew he would be capable of handling maybe 4 or 5 of those bottles in a day, but a quick estimate put this room at 400 to 500 bottles.

Jack looked around and found a small vial the same size as the ones he had found in the freezer. He opened one of the incubators and pulled out one of the large cell growing vials and took it to an isolation hood. He carefully pulled out a small amount of the liquid and then replaced the large vial in the incubator. Just to be careful, Jack sprayed down his hands and the small vial with 70% ethanol. That would be enough to kill anything that was on the outside of the vial. Then he placed the vial in his small cell freezer rig with the other vials.

Jack thought he had enough to work with now. He would have liked to have obtained a sample of the anti-viral drug so that he could see how it worked, but he wasn't going to press his luck. Just as he turned to leave, he heard voices in the anteroom. He looked up and saw two cameras in the back corners of this room.

Were they here to get him? He walked up to the door and heard "…name is Jack Griffin."

They knew he was here. More importantly, they knew his real name. But how could they know that? Jack looked around, but of course there wouldn't be any windows in a BSL-3 room. Jack had to find some place to hide until he could get out again. Now, especially, he was thinking that he shouldn't have come here. His heart was starting to pound so loud that his ears were ringing and his hands were shaking. He found a small closet for supplies along the same wall as the door, but they would probably look there first. There was a small space under each hood, but the hoods closest to the door would be visible from the ends. He couldn't get caught now. He had to make it to the outside door, and then he could run. He wished Ace were here with him. Who would take care of Ace if anything happened to him he wondered? His mind was wandering, and he needed to focus.

Jack noticed a small gap behind the incubators where all the tubes and electrical outlets were located, but he was skinny and could slip behind them. He chose the one closest to the door so he could try to slip out once they were in the room. As he passed behind the incubator, he found a large wrench on top that was used to change the CO_2 tanks. He grabbed this for a hefty weapon just as two men walked through the door decked out in the gown and other protective gear. Jack couldn't see their faces clearly enough to determine if he knew them.

"Bernie said he was in here. You go that way, I'll go this way. Saresh said he wants to talk to him, so don't hurt him too bad."

The speaker stepped in Jack's direction and slowly walked past the incubators and isolation hoods, looking back and forth. He walked past Jack and down the aisle. When he had walked about 15 feet past Jack's hiding spot, he slipped out the other side of the incubator and quietly made his way to the door.

"Hey, there he is! At the door."

Jack froze for an instant. This was how rabbits got killed, blinded by the fear of their attackers. That split second of a vision of a rabbit freezing still while a predator sneaks up on it, was enough for Jack to move again. He jumped through the door into the anteroom with their footsteps right behind him. He looked around the room and found a broom in the corner. He wedged the handle against the door and the other end against the sink in the anteroom. This would hold for a while.

"Hey! Hey! Bernie!"

Jack could hear them yelling, but he didn't wait around to take orders. Without removing his protective gear, he pulled open the door to the hallway and bowled over Bernie who was as shocked to see Jack as he was to see him. Jack kept his balance but Bernie hit the far wall hard. Not looking back to see how Bernie was,

Jack raced down the hallway to the large open laboratory, and the only doors he knew of.

This was stupid; Jack knew he wasn't a secret agent. He never should have come here. He should have stayed home with Ace. Amy was right; he didn't know what he was doing. Jack turned the corner after entering the laboratory and saw Aded just coming through the door by the security office. Jack stopped cold in his tracks stumbling backwards. Bernie was behind him, and Aded was in front of him. That left only one door remaining, the loading dock. Jack ran around the laboratory benches toward the back door with Aded right behind him. He could hear the thumping of Aded's boots on the hard concrete floor and knew he only had a second or two lead.

Jack reached the back door and swiped his ID badge over the door as he pulled on the handle. The door didn't budge, and he only had about one more second. He turned and saw in Aded's right hand a small black gun. Jack swiped the ID badge again and pulled on the door. The door opened this time and Jack jumped through the opening and down off the loading dock ledge.

The night had cooled since he had entered Intech, but with the extra protective gear on, Jack didn't notice the weather. Fear and running were making him sweat. As he ran, he stripped off the booties and other protective gear leaving a trail of clothing along the street that looked as though he was sprinting to his lover. His car was only two blocks away, but Aded was right behind him and he could hear the shouts of other guards coming through the loading dock door. He wouldn't have time to get to his car, unlock the door and start the car before they got to him. He would have to lose his attackers in the labyrinth of alleys around the old warehouses.

Jack turned right down one alley only big enough for personal access and raced along the building. The moon was bright enough to pick out the debris that had been strewn across the street allowing him to dodge the larger items and jump over the smaller ones. Jack ran as hard as he could taking as many turns as possible. Eventually he felt a distance begin to grow between himself and Aded. After six or seven blocks and countless turns, Jack stole a glance back over his shoulder. Aded was just turning the last corner almost a block away. He still held his gun out and was now pulling it up into an aiming position. Jack raced down the next alley and the next and the next for another five blocks before looking back again. He didn't see Aded this time.

Jack stopped and listened. He could hear some shouts, but they sounded far in the distance. He looked around and realized he had no idea where he was anymore. He was pretty sure his car was to the right and behind him, but he would have to wander around to try to find it...back the direction of Aded and the shouts. Jack looked around and saw a stair leading up to the door of a long abandoned warehouse. The door was a simple metal door with no windows, but someone long ago had forced it open so that, now with some effort, Jack could let himself in. He entered a building that was almost completely dark. The only lights were thin shafts of moonlight that filtered down through the dust from windows high up near the roof. Jack's eyes adjusted to the darkness of the building, and he saw great heaps of carpet piled into the center of the room. The smell implied that water had long ago seeped into the carpet beginning a mold process that would certainly end in a complete disintegration of the carpet mound. He also smelled the overpowering and distinct smell of ammonia. Urine. This was probably a favorite hiding place for homeless people, especially on a cool night like this.

Jack picked his way around the carpet mound toward the back of the warehouse. He found a small room in the back corner that had probably belonged to a foreman or manager. The door was open so he entered and saw a small metal desk near the back of the room and a pile of carpet along one wall.

"Hey, get otta here. Thish ish mine. Find yer own." A drunk man who had been hidden in the corner of the office staggered to a standing position.

Jack didn't say anything as he backed out of the room. He would try to find a place in this building to hide until he could get back to his car. He had no idea how many people might be in here so he would have to be careful. Jack sat down on a dry piece of carpet placing the large mound in the middle of the room between him and the door he had entered through.

Chapter 22:

Unable to see his watch in the darkness, Jack had no idea how long he sat there, but it had to have been at least two hours. He hadn't heard any more voices since he entered the building, and fortunately, he hadn't run across anyone else in this warehouse. The smell of ammonia from the urine was starting to overcome him, making him think that the carpet pile he was leaning against may be a community toilet for the building. Jack stood and stretched his legs and at that point noticed a searing pain down his right leg. He reached down and felt a large tear in his pants starting at the knee. Now he remembered hitting his knee on the door of the loading dock as he escaped Intech. He could feel a large scab beginning to form. He would have to treat that when he got home.

Jack made his way out of the building into a light that was so bright he thought for a moment he had slept until sunrise. The full moon was directly overhead now and clearly lit his path as he tried to find his way back to his Jeep. He walked slowly, partly to ease the pain in his leg and partly to conceal the noise of his footfalls in case anyone was close by. Every intersection he came to, he stopped and listened for any human noise. He also peered down all three other streets until he was sure there was no one watching.

It took him almost an hour to find his Jeep. He leaned up against a building and watched his car through a small view that was offered between a building wall and a telephone pole. He watched and listened for at least ten minutes but didn't hear or see anyone. Jack pulled out his keys and quickly made his way to the Jeep door, unlocking it as he pulled it open. He started the car and drove away with the lights off for the first two blocks until he reached the main street. He hadn't seen anyone in his rear-view mirror. Maybe they had given up on him.

After he was several miles from the warehouse district, Jack stopped and took his time to remove the cell freezer containing the virus samples from his pocket. He placed this in the dry ice cooler he had brought along. The cell freezer had started to warm, but he guessed it hadn't gotten above −20°C at the most. The virus would still be okay at that temperature. He could store the virus at that temperature for at least a couple of days without losing too much.

Jack looked at the dash clock and was surprised to learn it was three o'clock in the morning. He drove the rest of the way home without incident, but as he drove down his street, he noticed a man standing on the sidewalk outside his apartment door. He remembered now that they had said his name at Intech. They knew who he was. They hadn't been out searching the alleyways for him. They had simply come back to Jack's apartment to wait for him. Jack drove past his apartment and parked around the corner where he couldn't be seen.

Samantha answered the phone on the first ring and didn't sound as though she had been asleep.

"Samantha, it's Jack."

"Jack, I've been so worried about you. Where are you? Don't come to your place. There's some man outside that Brandon says he knows from Intech."

"I've seen him. I'm parked around the corner. Brandon's there?"

"He's been here all evening waiting for you to get back."

"Samantha, someone told Aded that I was going to be at Intech tonight. They were waiting for me, and they knew my name. I'm not sure I can trust Brandon. Is he listening to this now?"

"No, he's in the bathroom right now. But I can't believe Brandon would set you up like this."

"I'm not sure what to believe right now. I need to get Ace out of there, and then I can head out to my cabin. I want you to come with me."

"Not without Brandon. I don't think he would have done this Jack. But I'll help you get Ace if I can."

"Okay, let me think. I know, I need you to call someone for me. I don't have the number with me, but you can look it up in the yellow pages."

Jack waited around the corner and watched the man sitting on the front step to his door. Samantha had offered to come out and wait with him, but he needed to think about what his next step was going to be. Once he got Ace he could go out to is cabin for a few days and try to think about his next move. He would store the viruses in his freezer where they would be safe for a few days. He could examine the virus when he got back to lab, whenever that would be.

Jack heard the familiar rumble from down the street. It sounded like thunder from his days in the Midwest, except this thunder rolled down the road. Peeking out the back window of the Jeep, Jack saw three Harleys roll by with Snake in the lead. He would take care of this without asking any questions. Jack walked to the corner in time to see the man on the steps running down the street. Snake was holding a pipe in his hand with the two other guys backing him up.

"Thanks, Snake," Jack yelled down the street.

"Jack, I don't know what you got yourself into, but do you want us to hang out here for awhile?"

"No, I'm gonna get out of town for a few days. But I would like you to find out about the Nissan I told you about. I saw it parked outside the Hoffman Institute last night, and I think it will probably be there waiting for me today. Just find out who it belongs to."

"No problem. Take care of yourself and next time call this number." Snake slipped a card to Jack. "It's my cell phone, and I've always got it with me. You got my old lady jealous having that woman call me."

The sun had begun to rise and shadows were slinking back to their corners of the world to wait the next night. With his attackers gone and the daylight encroaching on the world, Jack felt considerably safer. Suddenly, the events of the last night seemed more a dream than reality, although the lack of sleep probably wasn't helping any.

Jack let the dog out and then fed him before heading over to Samantha's. She grabbed him and pulled him into her apartment in a big hug.

"Jack, I'm so sorry you got into trouble over this. I just know Brandon didn't do this though."

"I need to get out of town for a few days until I think some of this has blown over. Snake is going to look into some things for me and maybe we can get some answers. When I get back, I'll look into those virus samples. I don't think it's safe for you here either, so I want you to come with me."

"Of course. But I want Brandon to come, too." Samantha was starting to tear up.

Jack looked past Samantha and saw Brandon sitting on the sofa watching the two of them. "I guess...I can keep an eye on him then."

"Jack, I swear I didn't tell anyone about you going to Intech. I need your help on this. I wouldn't have done that. You have to believe me."

Jack was in no mood for an argument. "Just grab a few things, and we can get on the road. I don't want to wait until someone gets the bright idea to come back."

Jack walked back to his apartment and put the virus samples in the freezer before packing some food in a cooler. He grabbed a few items of clothing and Ace's things and packed them all in the back of the Jeep. Last, he went to his bedroom closet and pulled up a chair. He took a deep breath before stepping up and reaching back behind the sleeping bag and tent to find the wooden box. He pulled it down and opened the lid. There was the Colt 45 that had been his dad's. He remembered how his dad had been so proud of that gun; showing it to all his friends. Jack wasn't a big fan of guns, but kept it after his father's death as a reminder of those days when just the two of them were out in the woods shooting at soda cans. Now it might actually come in handy as more than a reminder of days without worry.

Samantha and Brandon were waiting on his front step as he locked his front door. "Okay, let's get going. C'mon Ace, get in back. Samantha gets to ride in your seat today."

Chapter 23:

Jack drove east directly into the rising sun, leaving the city behind him. The mountains rose directly ahead. He had made this drive many times since moving here. He often came out to his cabin for the weekend to get away from the lab and the city. It wasn't Montana wilderness with the dense, lush trees and freely running water, but it reminded him a bit of home for the lack of civilization. Jack drove the freeway east for about an hour before taking an exit and heading north on a small country highway.

He was driving for about a half hour when he saw a silver Nissan 300ZX on the side of the road. Jack slowed down briefly to get a better look. "Brandon, is that Aded's car?"

Samantha turned in her seat to look at her brother, "He's out cold in the back seat. Do you think it is?"

Jack didn't have an answer, but it sure looked like the same car that had followed him over the past two weeks. But what would it be doing out here? Did they know about his cabin? Had they already been out to it? He hadn't told Brandon about the cabin until this morning and Samantha had been with him since then. Jack bristled at the idea that his escape might have been intruded on by these people. He decided, though, to calm Samantha's fears.

"Probably not. I'm just tired and paranoid. What are the chances, after all?"

Jack drove on in silence while Brandon and Ace slept in the back seat and Samantha stared out the passenger window as houses and side roads started to give way to open desert and cactus. This was a very sterile place in the world, and Jack didn't know how anyone would have seen this as a paradise in which to live. The ocotillo bush and Joshua trees were the tallest plants out here at only about six feet. Most of the ground was barren with only large boulders to break the landscape. The city they left behind was a truly artificial place in this part of the country. Water was shipped in from the mountains and beyond to build a fantasy world for the inhabitants. People living in the city could believe, as long as they didn't leave the city limits, that they were in a lush garden paradise. An Eden in Southern California.

A half hour beyond the Nissan sighting, Jack slowed the Jeep and took a right turn onto a road so small that most people would have driven right by without ever seeing it. A mile down the road a small sign with hand painted lettering was stuck into the ditch. It read "15 Miles to the Jack Shack". Jack always smiled when he saw that sign, knowing he was getting close.

"What was that sign?"

"I was bored one weekend and I had the B52's song "Love Shack" in my head, so I measured out 15 miles and put that sign in."

"So this is your love shack?" Samantha asked with a twinkle in her eye.

"No, no. I really did it out of boredom. Look around you. There's not exactly a lot to do out here. My closest neighbor, Bill Walden, is almost two miles from my place."

Jack slowed the Jeep as he came over a slight rise in the road. In the distance he could see the sun reflecting off something. It was still a ways from his place, but he had driven this road enough to know it like the back of his hand and he had never seen anything new out here. People just didn't move out here for pleasure and they never built a bright shiny new structure. He knew Bill had no plans to build anything new. He was almost 80 years old and living off his social security.

"I wonder what that is?" Jack thought out loud.

The remark caught Ace's attention as he stepped over Brandon to get a better look out the driver's side of the Jeep. Brandon sat up and pushed Ace off his lap.

"What's going on? Where are we?"

"We're almost there. I was just wondering what that new building is over there."

As they got closer, Jack noticed a chain link fence topped with barbed wire lined the property along the road. Set back about 100 yards from the road was a new galvanized steel shed about the size of a barn. He made a mental note to ask Bill who had moved in here and if they were planning to farm the land.

Jack drove on down the road leaving the new barn beyond view. About 10 miles farther Jack pulled into a dirt driveway marked only by a mailbox with the label "Jack's Place".

Another hundred yards down the road, Jack pulled up next to his cabin. Jack and Ace had been here many times so nothing was new to them, but Samantha and Brandon were seeing it for the first time, and the outward appearance could be quite deceiving.

"You're not the next Ted Kaczynski, are you?" Brandon asked.

"Trust me, the inside is a lot nicer. Just remember, this is only a cabin, and I do live on a post-doc salary. Also, I never expected to bring guests out here, so most of the cabin is set up for my tastes only."

"I think it has charm," Samantha said.

The one-story wood-sided cabin was built in the style of the California bungalow at one time, but over the years previous owners had drastically altered this small unimposing structure. The front and side of the house now had a wrap-around porch attached to it, which would have been charming, if not for the blue, yellow and white paint peeling from the wood. There were several styles of windows indicating the various additions the house had endured. The front windows had probably been the original wood-casing single-pane glass. To the side of the house by the driveway, were flush aluminum-clad windows, indicating an addition that was probably added in the 50's or 60's. At the back of the house was what appeared to be an enclosed porch with newer, probably 80's vinyl windows. Behind the house and peeking out around the corner was an old red barn that looked as though someone had spent more time and money making sure the barn was in better condition than the house.

Ace was barking at the scrub brush around the property, racing back and forth near the house and barking under the porch.

Samantha had been worried ever since Jack had seen the familiar silver car on the side of the road, "Is something wrong?"

Jack laughed as he watched Ace jump into a mesquite bush. The long day was finally catching up with him. "No, no." Jack tried to explain but had tears streaming down his face. "That crazy dog is trying to herd up the gecko lizards. Trust me, as long as he is out here, he will do all he can to make sure we are safe from the geckos."

Jack grabbed several bags from the back of the Jeep. "Ace, inside!"

Jack unlocked the front door and pushed it open with his foot to allow everyone to enter. To the right was a small fireplace that had obviously seen much use. Wood was still piled alongside the hearth. In front of the fireplace was a futon sofa that Jack used as a bed when he was here on his own. The only other furniture was a small table in the dining room beyond the futon. Samantha walked past the table and looked into the bedroom to the right and saw a queen size, wrought iron, four-poster bed that took up most of the whole room.

Beyond the dining room was the kitchen and from this view, Samantha could see that the addition on the side of the house had been an extension of the kitchen. Professional chefs would have been envious of this kitchen. A large, six burner Viking stove and Sub-Zero refrigerator dominated the left side of the kitchen. A butcher-block cutting board counter occupied the middle of the floor. The previous owner had obviously placed a great importance on cooking.

Beyond the kitchen and to the right was a small bathroom that was at least clean, if nothing more. The back addition had been completed for either an office or a second bedroom, but the principal feature of this room was the large full size window along the whole back wall facing east allowing the early morning sun to light the entire house.

Samantha turned back to the front of the house and saw Jack and Brandon standing just inside the door with Ace sitting at their feet. They had been watching her assessment of the house. "Yes, this will do," she said with a laugh.

"It's good to see you laugh again." Jack said.

"I guess I'll get the sofa," Brandon said, leaving an uncomfortable silence in the room.

Jack didn't want to get into sleeping arrangements yet, although Brandon's idea suited him just fine.

"I brought enough food for today and tomorrow morning. There's a small store about 10 miles farther down the state highway, so we can go there tomorrow."

"How long do you think we should stay here?" Samantha asked.

"I don't know. I don't know. I need to think about some things. Right now, I need a nap and some breakfast."

"Why don't you take a shower and then a nap, and I'll make bacon and eggs for all of us when you get up." Samantha said.

Chapter 24:

Jack felt better after his shower, but he truly felt refreshed after sleeping for three hours. The activity and stress of last night had hit him hard. He opened the door to the bedroom and was surprised by what he saw. While he slept, Samantha had cleaned his house and had cut some of the desert wildflowers and placed them throughout the house. She had washed all the windows so the room looked much brighter with all the new light coming in. Jack could already smell the bacon on the stove.

"You didn't have to do this."

"I know. I wanted to. If I'm going to be staying here for even a few days, I need a home, not a bachelor pad. I've been wondering what's in that barn out back. There is a large lock on the door."

"It's the garage. I'll show you later. Where's Brandon?"

"He took a nap on the couch and then went for a walk about ten minutes ago. He'll be back in a few minutes. He has this ability to sense when I am cooking and will arrive just as I am finishing so he doesn't have to help." Samantha winked at Jack.

"Okay, I get the hint. How can I help?"

Brandon did indeed show up just as Samantha was putting the scrambled eggs on the table. After breakfast, Jack and Samantha walked back to the garage to and left Brandon to wash the dishes. Jack unlocked the doors and pulled the two sliding doors open to allow sunlight into the barn. In the middle was a half assembled Corvette convertible. The top was gone and the doors and hood of the car were resting against one wall. Along the other wall was a large workbench with tools neatly hanging on hooks on the pegboard behind the bench. Bolted to a stand in front of the car was the large V8 engine.

"It's a Corvette?"

"Yes, it's a '67 Corvette convertible 427. It was my dad's favorite car. He talked about getting one his whole life. He bought this car a year before he died with the intention of restoring it. We never had much money, so he couldn't afford much. This car was such a wreck when he bought it. He said it would be a good learning experience to rebuild a car from the ground up. He had just started the project, and then he died. I brought the car down here, and on free weekends, I come out here to work on it. I want to finish it for him. It isn't even my favorite car. I would have picked something else."

Samantha walked around the car getting a good look at it. The color was faded from years in the sun, but she could tell it had been red at one time. The wheels were off the car now and were piled in the corner of the barn. The seats were white vinyl but had cracks and tears and were stained brown from the years and miles of passengers. The carpet inside the car had been ripped out, and rust was showing through in some spots of the floor.

"I like it. It has promise. I think it wants to be driven again. My dad always wanted a Corvette too, but he wanted one of those Stingrays from the '70's. I think there is something very American about driving a real muscle car."

Samantha walked to the back of the barn and found a couch against the back wall. "So, how did your Dad die?"

Jack walked to the back of the barn with her and looked closely at the engine mounted on the stand. "Pneumonia. Actually it was influenza. He had this rare genetic condition where his body didn't produce a certain type of antibody called IgA. That's the antibody that protects you against infections of the lungs. He knew infections as simple as a cold could be deadly to him, but he was too stubborn to see a doctor.

"He was sick for a couple of days, but he kept telling me it was just the flu, and he would get over it like he always did. I found him one morning at his bedroom doorway. He didn't have the strength to stand up. He looked at me with tears in his eyes. I had never seen my father cry before and it scared me. Looking back on it now, I know why he was crying. He knew. I took him right to the hospital, but he died a few days later. I was holding his hand when he took his last breaths.

"I was one year from finishing my doctorate. He never got to see me graduate. He never got to see me get my Ph.D."

"But he knew. He knew you were going to get it."

Jack started to tear up so he turned away from Samantha. "I never had a chance to show him what I could do with my life. I never got to tell him how much I had learned from him. I never got to…I never got to tell him how much he meant to me."

Jack seemed unaware that he was telling this to Samantha or even that someone else was there with him. He suddenly felt he needed to get this out. "That stubborn bastard!" Jack yelled, "I told him! I told him we should go to the hospital. I told him the flu could kill him.

"And now what? What am I supposed to do? He was always there for me. He helped me decide what college to go to, what motorcycle to buy, what…

"I never told him I loved him. I never…I never…I never told him I would miss him." Jack voice broke. "I do miss him. I miss calling him and asking him some stupid question just so I could talk. I miss showing him the motorcycle I finished. I miss telling him about my new job. I miss having him meet Ace." Jack smiled about that last one.

Samantha stood up and grabbed Jack around the shoulders spinning him to her. She pulled his head into her chest as Jack let loose and cried.

"Sometimes, I just feel alone. I come out here and I want to work on this car, but I just sit and think about what it would have been like to work on it with him. That was his plan."

Samantha didn't say anything, guessing that this was a time to just let Jack talk.

"I wish he was here, so I could ask him what to do now."

Samantha whispered into Jack's ear, "He is here now. He'll always be with you. You don't need to ask him what to do, you already know what he would say. Don't you see, that's what he taught you."

Jack looked up into Samantha's face. She wiped the tears from his cheek with her thumb and then leaned down and kissed him tenderly on the lips. Suddenly, Jack didn't want to think about his dad anymore. He didn't want to think about this car. And he didn't want to think about the trouble he was in back in the city. He just wanted to be here with Samantha. Samantha pulled her head back and Jack leaned in to kiss her. She opened her mouth as he kissed her upper lip and then he kissed her more deeply, passionately.

Jack wrapped his arms around her waist feeling her hips beneath her jeans and pulled her to him closer. He kissed her again and felt her hands rubbing across his back. He felt her reach down and undo his belt. He let his hands drift down to her hips and then to her thighs. He pulled her up so that he was carrying her. Their lips never parted. Jack knew this was something they both needed now. A moment to give in to a guilty pleasure during a time when they both knew difficult decisions needed to be made and difficult choices were coming up.

Jack gently put Samantha down and went to close the barn doors to give them some privacy. Samantha moved back to the couch and was lying down undoing her pants and pulling her shirt out. Jack walked back and lay down next to her letting his hands explore her whole body.

He had never wanted to be with someone more than he wanted to be with Samantha now. He helped her off with her pants as she pulled her shirt off. Jack pulled his pants off and lowered himself down to her caressing her face, thinking he must be the luckiest person to be here now with such a beautiful woman. Right now, he didn't think anything could feel better than this.

Later that night the three of them had a nice dinner of spaghetti with homemade meatballs. Jack had brought along a bottle of Merlot that they all shared. It had been a long night and day for all of them. Even with his nap, Jack was ready to turn in early. Fortunately, there were no awkward moments for sleeping arrangements that night. Jack had spent many nights in this cabin, most sleeping on the sofa near the fireplace to keep warm, but a few in this bed. However, this was the first time he slept here and felt truly at home. There was something right about sleeping next to Samantha. They seemed to fit together.

Chapter 25:

Jack woke up at first light and found Ace resting his head on the bed next to him. As was usually the case, Ace was so excited to be out of the city that he had to get outside as soon as he could. Jack put some shoes on and let Ace out the back door. He walked along the back of the property with no destination in mind. The sun was rising over the plains to the east reflecting off the mountains to the west painting a brilliant red and orange landscape.

Ace continued his dogged pursuit of the geckos hiding in the scrub brush as Jack walked on. He really liked the way things were progressing with Samantha. He hadn't told her yet, but he thought he was falling in love with her. Last night before they fell asleep they spent an hour telling stories from their childhood. Jack learned that she and Brandon had never really been close while growing up. He was three years younger than she was, and all through school she had felt he was more a pain than a blessing. He was always following his older sister around even through high school. With Brandon getting into trouble all the time, she felt she was constantly bailing him out and trying to stick up for him even when she knew he had been in the wrong.

It wasn't until she left for college that she realized she really did have a close bond with her younger brother. All those years of doing things together, just the two of them, had formed a relationship that only two siblings can know. Unfortunately, it was then that he started rebelling against his parents. And it was then that he started to pull away from his sister with his secrets.

Jack hadn't realized it but he had walked beyond the border of his property and was heading toward Bill Walden's place. Ace was leading the way, running back and forth clearing the path of lizards for his owner's safety. Bill was outside working on his truck when Jack walked up.

"Hi Jack. Good to see you. Haven't seen you 'round much lately."

Bill was a tall, thin black man who had been raised in Louisiana, although he pronounced it "Loosanna". He had never attended school but learned farming from his father. He moved out to Southern California with his wife and three sons in the 70's to plant an orange grove. Back then, the state was all too happy to divert water from the few rivers to irrigate the citrus groves. Things changed as the population ballooned and the state started funneling more water to the large Southern California cities and started charging more for the water. Bill couldn't get enough money for his oranges to pay the bills so the state bought his water rights from him. He had retired on that money. Bill's wife died a few years before Jack bought his land, and his sons almost never came to visit.

"I've been busy at work. Haven't had time to get out and work on the car."

"How's it comin'?"

"Slow, slow. I might have the engine done with a few more weekends. I might need your help putting it back in the car though."

"Sure. Oh, where's my manners. Wanna a cup of coffee?"

Jack followed Bill into his house. "Hey Bill, do you know what that new barn is for down the road?"

"Ain't no barn as I've ever seen. And they don't have no animals or equipment for farmin'. No, I don't like what's goin' on down there I tell ya."

"What do you mean?"

"They put up that big fence all 'round their place. And when I went down to meet the new neighbors, they had guys with guns standing at the driveway. They were real tanned guys, not black like me, but dark skin. An' they got an airplane too. Lands right there on the other side of that barn. No, I don't like what's goin' on down there."

"Really? What did they say when you went up to the guards?"

"They jus' said this was private property and I should mind my own business like they own this whole area. I tell ya, things have really gone down 'round here the past few years. Used to be that neighbors would talk to each other. I tell ya somethin' else too, they had a weird accent. And they're always doing stuff at night there too. Not during the day like normal folks."

"Maybe I'll try to check out the place. I'll let you know what I find if I do. Thanks for the coffee Bill, but now I should get back to my place."

"So soon? I could make eggs for ya. They're fresh from the hen this morning I tell ya."

"Sounds great, but I have some friends at my place this weekend. I should get back to them."

"Well, take some eggs back with you for breakfast. I've got too many."

When Jack arrived back at his place, Brandon was up and reading an old car magazine of Jack's. Samantha was just getting out of the shower. "I've got eggs, just give me a second here and I'll have breakfast for all of us."

"What? Do you have chickens here?" Brandon asked.

"No, I was just down talking to my neighbor Bill. He has chickens."

"Jack," Brandon came out to the kitchen and started helping Jack prepare breakfast. "I want to thank you for all your help lately. I'm so sorry you ran into trouble at Intech, but I swear I didn't tell anyone."

Jack stopped cooking for a second and looked Brandon in the eyes. He had to admit, he couldn't find much reason for Brandon stealing an ID badge, giving Jack all the information he needed to get in and then letting Aded know he was going to be there. Deep down, he believed Brandon, but he would still keep an eye on him. "I believe you, Brandon, but that is just one problem now. They know where you, your sister and I live. More importantly, they know that I was in Intech and they don't know what I know which makes me a wild card in their eyes. We may have to face the possibility of going to the police."

"No! I won't put my family in that kind of danger. These guys have people all over."

Samantha came out of the bathroom with her hair tied up in a towel. "What are you two arguing about?"

Brandon gave Jack a hard look, "Nothing, Sam, just talking about where Jack got the eggs."

"Well, I was thinking about heading into town to get some more food after breakfast. How long do you plan to be out here Jack?"

"I'm not sure yet. Just get enough for a couple of days, and we can figure out what to do after that. I'll stay here and get some work done around the yard."

Chapter 26:

Samantha drove the Jeep and took Brandon with her so that Jack could have some time alone. He had to figure out what to do next with the virus samples. Brandon was refusing to go to the police, which put him in an awkward position since his own life was also in danger. He was sure that if he went back to his apartment, Aded would be waiting for him.

Jack grabbed the Frisbee off the counter and walked out back with Ace. He threw the disk several times, watching as the dog ran back with his toy, so full of happiness. Jack knew the dog really enjoyed coming to the cabin because he got to spend so much time with his owner. Jack sat on the back step and brushed Ace's fur while he thought about things.

He wanted to do as much as he could to help Samantha, but he had no idea what his next move was going to be. Just a week ago, he had been happy in his routine of working with Amy in the lab and walking Ace in the evenings. On an occasional weekend, he would head to the cabin and work on the car to take his mind off the science.

He had purchased the cabin as an escape from the city. Growing up in Montana, Jack had spent many hours with his dad camping in the woods or fishing the area streams. At the time, he had seen it as a chore to spend time with his father when his friends were going to the big city to spend days at the amusement park, but looking back on it now, he realized that he had really cherished those times. He had bought this place and the land from the money his father had left him. He knew it was a place his father would have liked if he had been forced to live in a big city in Southern California.

Jack must have dozed off for a few minutes, but he woke when he heard the Jeep drive in the driveway. Samantha came around the back of the house and saw him sitting on the step.

"I asked Brandon to put away the food so we could talk. He told me you wanted to go to the police."

Jack got up and started walking toward the back of the property with Samantha beside him. "I just don't know what else to do. I don't know how to handle danger like this. I'm just a scientist."

"Jack, you've done so much for us already. Don't you see, you're the one who volunteered to follow Brandon? And you're the one who went to Intech to get some of the virus samples. You told me earlier you wanted to change the world. This could be your chance."

"I wanted to change the world, but I didn't want to be a hero for it."

"But you are a hero. Some people spend their whole lives striving to be heroes, while others are destined to be heroes. They spend their whole lives doing what comes naturally until some day someone tells them they've been heroes all along. You're that kind of person Jack. You've been a hero since I first met you, and my guess is you've been a hero for longer than that. And all this time, you've just walked through life doing the right thing when the time came. I've never known another person that would have taken the risks you took for Brandon and me. Jack, you can change the world, you can do it as a scientist, but first you need to accept who you are. Your dad would have been so proud of you. You are going to change the world one day, but right now, you're changing our little part of it."

Jack was thinking about what she had said. He hadn't really thought that he was taking risks, but he was. He always thought he wanted to live his life in the lab and at home with his dog, but lately, he had found himself doing a lot of things he never would have thought he would do.

He already knew what he needed to do next. He guessed he knew even as he was running from Intech. One day he wanted to spend a day with Samantha when all they thought about was the warm sun beating down on them and the clean, saltwater air they were breathing. There was something he needed to do before he and Samantha would be able to enjoy that day.

"I need to get back in the lab without Aded and his men seeing me. And I have to work without a lot of other people around so I can avoid questions. I guess I could work in the evenings, but I would have to stay near the institute. I could stay at a hotel and work at nights." Jack had been thinking out loud but now looked at Samantha and saw the worried expression on her face.

"I know you need to do this, and I thank you for doing this for Brandon and me, but I'm worried about you going back there."

"I think you two should stay here while I'm gone. It will be safer. I don't know how much they know about Brandon helping me."

"Be careful Jack. I'm getting used to having you around."

Jack stepped toward Samantha and held her by the waist, kissing her gently on the lips. "I promise I will be back for you."

"You'd better."

"I don't know how long this will take me. If everything goes well, I would bet a few days to a week."

"That's okay. When we were in town, I called into the paper and told them I had a family emergency and would need the week off. I also picked up some paint for that porch of yours. Brandon and I will earn our keep around here."

Jack smiled this place definitely could use a woman's touch. "I want to stay here tonight with you, and then I'll leave tomorrow afternoon."

Chapter 27:

The rest of the day Jack and Brandon worked on the Corvette engine while Samantha baked a large farewell dinner for Jack. Brandon turned out to be quite adept at engine work, and he and Jack were able to get most of the rest of the engine together. With Brandon's help, Jack knew he would have been able to start assembling the car by the end of the week.

Samantha cooked a honey-glazed ham with homemade corn bread muffins. She and Brandon had picked up some wine while they were in town so they were all able to enjoy a big meal with a bottle of Pino Grigio.

That night they huddled around the fireplace and talked. Although days in the desert during November could still get into the 90's, the temperature at night fell as low as the 40's. Jack explained that he planned to stop at his house and pick up the virus samples he had stored in his freezer. The few days they were at $-20°$ Celsius, which is the temperature of a normal freezer, wouldn't affect the virus too much. He would try to grow the virus on cells that he knew would support the growth of influenza. Then he would try to sequence the genome so that he could determine exactly how infectious and dangerous the virus was.

"Isn't it dangerous working around viruses like this?" Brandon asked.

"Well, I normally work with the Adenovirus, which is just the cold virus. The one from Intech is influenza, which causes the flu. The thing is, even though influenza can be deadly to some, to a healthy adult like me, it really poses no threat. Both of these viruses are worked with in a BSL2 facility which means that I have to wear a lab coat, gloves and work inside an isolation hood."

"You mean one of those things where you put your arms inside of gloves to work in a box?"

"No, it's called a laminar flow hood. It just means that air is constantly circulated in such a manner to keep a barrier between the inside of the hood and the outside air. There is a space about a foot high at the front where we can put our arms into and pass materials in an out. The circulated air is filtered through a HEPA filter which stands for High Efficiency Particulate Air filter. So it can filter out aerosolized viruses and bacteria. Then, we autoclave all the waste from the hood. It's really easy to work in."

"Maybe when this is over, do you think I can come in and see the lab?" Brandon asked.

"Sure, I love showing people around the lab. I think it takes some of that Dr. Frankenstein working in his ivory tower status off of us scientists. A lot of press these days focuses on the evil that we have the potential to do, but in reality, we're just people trying to earn a paycheck like everyone else. The paycheck is pretty small for what we do, so most scientists picked this field for the opportunity to help people. I can honestly say I don't know a single person that would like to clone humans, or release a super virus on the population or design gene therapy for altering muscle or intelligence."

The rest of the evening, the talk moved toward the restoration of the Corvette and Jack's work on the Panhead Harley he owned.

The next day, Samantha and Brandon went into town to pick up supplies for the whole week. They bought groceries and books to read and Brandon even picked up a shop manual on the '67 Corvette. Jack walked down to Bill's place and explained that his friends were going to stay at his place but wouldn't have a car, so they would have to come down to him if there was an emergency. Bill seemed to relish the idea that he might actually have some visitors and immediately began cleaning his place.

Chapter 28:

That evening Jack kissed Samantha goodbye and left the two of them and Ace at his place as he drove back into town. He arrived back at his apartment well before the sun set and too early to try to sneak into his place. Jack parked around the corner and scouted out the cars in front of his place looking for anything suspicious. He didn't see the 300ZX, but then again, that would have been too easy. Jack examined all the cars on the street without seeing any that stood out as not belonging here. He looked in the windshield of as many as he could but didn't see anyone. He decided to wait until dark and then check around again.

The sunset earlier now that daylight savings time had ended and the Earth was approaching the winter solstice. Streetlights blinked to life at 5:00 and the last bits of sun faded by 5:30. Jack waited another hour before walking along the sidewalk across the street from his apartment. He didn't see anyone suspicious and there didn't appear to be much foot traffic in front of his place. Jack took the bold step and crossed the street to his side and walked to his apartment quickly opening the door and slipping inside. He waited for his eyes to adjust so that he wouldn't have to turn on a light. Briefly surveying his apartment, he became assured that no one was in there with him. He went to the freezer and pulled out the virus samples he stole from Intech and packed them into a cooler with ice. As he left his apartment he looked up and down the street, but saw no one. Jack felt a bit foolish for acting so clandestine when it was now apparent that no one was waiting for him. Maybe all this was for nothing.

Most evenings, the lab was quite empty, and tonight was no exception. Only Jason was in the lab tonight and he appeared to be absorbed in a computer game he was playing over the institute's network.

Jack pulled the virus samples out of the package, freezing two vials and taking the other five for his study. He looked in the incubator and found that Amy had the cell line that he needed to grow influenza. A virus typically caused a specific disease because of the area of body it infected. The cold virus produced a stuffed up nose because it infected the nasal passages, HIV weakened your immune system because it infected the immune cells and influenza infections resulted in chest congestion because it infected the lungs. Since viruses were typically highly specific in there sites of infection, this required Jack to grow the virus on lung cells. Fortunately, Amy had some of these cells growing. He would replace them for Amy when he had a chance.

Jack diluted the viruses and dropped some of each on a different plate of cells. He would check this in a day to see if there was any effect. The virus infection should cause the cells to look sick or different by the next day, and within two days, most of the cells would start to die as the virus replicated. As a control, Jack dropped some of the virus sample on the cells he kept for growing the Adenovirus he worked on. The influenza virus shouldn't do anything to these cells.

Next Jack took a sample of each virus to have the genome sequenced. The influenza genome only had a few genes compared to the thousands that were in the human genome, but Jack only needed to sequence two of these. The rest of the genes were similar across all the different strains of the virus, but the two genes that were on the outside of the virus and were involved in the actual infection, were the genes that varied the most. If these genes infected lung cells inefficiently, the virus would cause only a mild disease; however, if they infected the cells efficiently, the virus could be deadly. These were also the two genes that varied most between the influenza A and influenza B. Jack would be able to determine a lot about the virus from knowing these sequences.

The problem with sequencing a gene was he needed to know a small part of the sequence to start. This would serve as the start or primer for the reaction. Then he would know the order of the sequence when he got it. Fortunately, the influenza genes all started the same way, so Jack could use that sequence to start the reaction. He would get those results the next day too.

The next step in studying the virus was to try to infect mice. Although mice typically didn't get the same diseases as humans, they could serve as a useful model for certain aspects of virus replication. Some strains of mice showed only very mild symptoms when infected with the influenza virus while other strains were much more susceptible to the virus. Looking around the room, Jack noted that the only strain he had available to him were mice that would show only mild symptoms. He would have preferred a more susceptible strain in order to determine the virulence of the strain he had, but he would have to work with what he had.

Jack diluted a small amount of the virus with normal saline solution and transferred this to an aerosol container. The mice were placed into a small plastic box to be anesthetized by chloroform gas. Within minutes, they began to sway and finally fell down to sleep. Jack sprayed the virus solution into the nose of one mouse and then returned it to its cage. This process was repeated for the other four virus samples.

Unlike in the movies, Jack knew that real science took time. He had done about all he could today and would have to wait until tomorrow for even the first of his results. It was only midnight now, so Jack spent the rest of the evening working on his anti-viral project. He hoped Amy had achieved some success with his work while he had been gone, but he would have to wait to ask her. She would probably be working late tomorrow night so he could ask her then. He looked forward to telling her everything that happened to him at Intech.

At 6:00 AM Jack was exhausted. It would take him a while to get used to this third shift work at the lab. He left then to avoid the questions of co-workers as they arrived for their normal working hours. He also wanted to avoid Dr. O'Leary for now, and he knew the professor usually arrived at 8:00. Jack drove north along the coast and then headed inland to a small town off the usual tourist circuit.

Jack looked for hotels that were off the main road feeling these would probably be the cheapest. The hotel he chose had a large sign proclaiming that this was the City Motel. The parking lot wrapped around from the office to a long single story building with several doors for individual rooms. Only two other cars were in the parking lot making the Vacancy sign unnecessary. At $25 a night, even a post-doc could stay at this place, although Jack knew he wouldn't want to have to stay too long.

The next evening, Jack arrived at the lab at 8:00. He was surprised to find Amy wasn't there anymore, but he did see another lab mate who told him Amy had been sick the past few days. That would explain why he hadn't received any of his data recently.

Jack checked for his virus DNA sequence, but was surprised to find a message from the sequencing center saying they were unable to sequence the virus with the primer he had provided. He had carefully chosen a sequence that should have been conserved across all strains of influenza. He sent the virus in again with a different conserved sequence.

Next, Jack checked his lung cell infections, but was surprised to find few if any sick cells. Maybe this virus was very slow growing and would take a few days to show signs of infection. If that was the case, then this virus might not be very deadly at all, and all this worry would have been for nothing. He knew that it would be a few days before the mice would show signs of infection, so he didn't check them now.

Once again, Jack was left with little to do for the night while he waited for data on these experiments. Hopefully tomorrow night he would get the results he wanted, and then he could head back out to the cabin. In the mean time he would just have to work on his normal project.

Chapter 29:

The next evening Jack was glad to see Amy back in the lab and working late. "Amy, you look awful. You shouldn't be working this hard if you're sick."

"Jack, I'm so glad to see you. I wondered what happened to you."

Jack didn't want to tell her about his troubles at Intech yet. "I went out to my cabin for a few days. I was hoping to get a little help from you, but I can see you're pretty sick."

"I'm going to be here for a bit longer today, and then I'll go home to sleep. Let's go get something to eat at the vending machine."

"So I got a sample of the virus from Intech."

Amy stopped at the vending machine. "Really? And you have it here?"

"Yeah. I was trying to sequence it. I used the common sequence from all influenza viruses, but it came back with nothing. I even tried another sequence for a primer, but today I got another note saying they couldn't get any sequence."

"Did you try infecting the lung cells I was growing?"

"Yeah, but that was only two days ago. Yesterday, there wasn't anything to see. I haven't had a chance to check them yet today."

"Well, let's go see what they look like. Are you sure you have influenza?"

"Brandon was sure that was what Aded told him Intech had."

Jack followed Amy into the cell culture room and placed the dish of cells under the microscope motioning for Amy to have a seat. "So, here are the lung cells."

"Yes, these cells look just fine. I can hardly see any sick cells." Amy sat back from the microscope. "Did you try any other cells. I mean just to see."

Jack reached back into the incubator and pulled out the nasal cells he used to grow the Adenovirus. "I put a sample on these cells just as a control infection." He handed the dish to Amy.

Amy looked at the cells under the microscope for along time, moving the dish around to get a look at as many as possible, "Jack, these cells are all dead."

Jack sat down at the microscope to get a better look. "Holy mackerel! Maybe I should be trying to sequence for the Adenovirus genome."

"I thought Adenovirus wasn't deadly."

"It generally isn't when it infects a human or animal, but in that case you have an immune response to fight it off. In this case, we are directly infecting the cells without any immune system. The virus usually will kill these cells.

"I'll send the virus back down to the sequencing center and have them try again looking for Adenovirus sequences. If all Intech has is the cold virus, we don't have too much to worry about. It's still highly unethical to spread the virus for the purpose of testing a new anti-viral drug, but at least it won't be deadly to people."

Jack and Amy walked back to the lab and sat at a small conference table to eat their chips and drink their soda, the usual dinner diet of the scientist.

"So did you have any trouble getting the virus from Intech?"

"Well, I did run into some trouble. But it wasn't much," Jack lied. "Just to be on the safe side, I want to work on this only at night when no one is around to ask questions. Can we keep this just between the two of us?"

Jack returned to the hotel the next morning for what he hoped would be the last time. The cheap curtains on the window performed a poor job of preventing the sun from entering, thwarting his attempts at daytime sleeping. To make it worse, the shower only dribbled water that was luke-warm at best.

Chapter 30:

Jack arrived at the lab at 10:00 that night, and of course Amy was waiting for him. She looked much better this night.

"I came in late today so that I could stay late and help you. I'm curious to find out your results."

"I just got the sequence back from the lab and I was going to compare it to known sequences in the database."

"So it is Adenovirus?"

"Well the fact that we got sequence information from the Adenovirus primer seems to say that. Now if we can just see what strain it is. I'll just go to the NIH website and enter our unknown sequence here, and then wait for the computer to spit out comparisons." The whole operation only took 30 seconds. Even during the busy part of the day, the National Institute of Health computers could do an operation like this in a few minutes.

"Okay, it says here that the first part of the sequence matches the Adenovirus strain Ad5. I know this strain; it's commonly used in the laboratory and doesn't really cause disease. There are a few discrepancies here and here." Jack pointed to spots on the screen where the sequence he entered and the known virus sequence didn't match up. "Viruses are known to make mistakes when copying their genome. It's sort of a built in evolution process for viruses. A few differences from the known sequence wouldn't be that unusual, but these are a bit longer than I would have expected. And then down the sequence a bit further, here, is a gene that doesn't belong in the virus."

"What is it?"

"Well, the program was trying to match our whole sequence against another whole sequence, so it just removed that gene to do to the comparison. Let's try just entering that unknown sequence into the database and see what we get back." The results were even faster this time because Jack was putting in a smaller sequence. "This must be some mistake. We must have contaminated our sample."

"What is it?"

"Well first we get back a laboratory strain Adenovirus which isn't too startling, but now this unknown gene is coming back as ricin toxin. Except that once again there are a few changes in the gene. Do you recognize these changes?"

"No. We could have the virus sequenced again. Maybe it is a mistake."

"Yeah, yeah." Jack was lost in his thoughts, picturing the virus the sequence was trying to describe to him. The Adenovirus was a cold virus that was easily spread among populations. It began by infecting the nasal passages and then moving to the upper respiratory tract. An infected person spread the virus easily by coughing or touching a surface after touching their nose. Adenovirus was stable in the air and on surfaces such as doorknobs and desk surfaces. The virus he had just received the sequence of would spread as rapidly as the common cold but would harbor a far more deadly secret than merely sniffles and a small cough. This virus was engineered to be a lethal weapon that once initiated would continue to spread through the population like a nuclear chain reaction. If this was true, that would mean he had one of the most deadly weapons man had ever dreamed of in his most horrific nightmares. Suddenly, he remembered the mice he had infected.

Jack jumped up from the computer, grabbing his lab coat while running down the hall to the animal room. Animals were kept in an isolated part of the building to avoid any spread of disease. Most people thought this was to protect the scientists, but really it was to protect the animals. One mouse strain could be worth tens of thousands of dollars. A lab like Dr. O'Leary's would have probably ten to fifteen different strains, and that was just one lab in the institute.

Jack ran through the double doors and stopped to slip on booties, a hair net, face mask and gloves before entering the mouse room. His mouse cages were gone. He looked around the room before finally going to the animal care staff station where he saw a note.

'Cages 13.1 and 13.2 all mice found dead Tuesday A.M., November 18'

The mice had died less than twelve hours after he had infected them with the virus sample. Jack felt a sick feeling in his stomach. Suddenly he no longer thought the sequence data was a result of contamination. His mice had died 12 hours after being inoculated. His cells had died a full 48 hours sooner than cells normally infected with Adenovirus.

He had been carrying that virus around with him. He had put it in his freezer and had worked with it in a BSL2 room. That was it! That was why Intech had such a large BSL3 facility. Influenza and Adenovirus alone didn't require a BSL3 facility. Neither virus was known to be deadly to a healthy adult. But an Adenovirus with a toxin gene in it would have been deadly to humans and would have required at least a BSL3 room if not a BSL4 room where everyone would have been wearing a full head to toe protective suit with an isolated air supply. Jack knew there were only two of those facilities in the United States and Intech never would have received a permit to build one. So they must have just insisted on care being taken and used a BSL3 room.

Jack realized now he had uncovered something much bigger than he could have ever realized. Aded didn't want to stop him from stealing the influenza virus. He wanted to stop him from finding out about their deadly virus. He had to get back and talk to Amy now. He had to figure out whom to tell, and he had to get back to Samantha and Brandon. There was no avoiding going to the authorities now. Millions would die if Intech released this and let the ocean breezes blow it over the city.

Chapter 31:

Jack ran down the hallway and into the computer room where he found Amy still at the computer. Jack was surprised to see the concern he felt mirrored on Amy's face.

"What, what is it?"

"I thought I would look at those mutations you found in the ricin toxin gene," Amy said with a quiver in her voice. "I searched against published articles on the toxin, and I found this paper that described these mutations."

Jack looked at the screen and saw the title of the paper Amy had pulled up.

Selected Mutations in the Ricin Gene Increase Toxicity and Stability

"So the virus carries a toxin gene that is even more potent than the natural gene. This is getting worse. But that would explain why the mice died so quickly."

"The mice are dead already?"

"They were dead the morning after I infected them."

"This is bad Jack. Look here too. The first author on the paper is Dr. O'Leary."

Jack stared at the screen. The paper was published in 1978 so that would have been about the time that Dr. O'Leary was starting out here at the Hoffman Institute. Jack looked over the paper, looking at the figures which showed an almost ten fold increase in the ability of the modified toxin to kill cells. Dr. O'Leary had done experiments on mice and showed that when the toxin was inhaled, it killed mice three times faster and with a dose five times less than the normal toxin. Jack scrolled down the screen to the end of the paper and read the acknowledgments section.

This project was funded by a grant from DARPA.

"Amy, did you see this?" Jack pointed to that last line of the paper. He had a hunch, "Why don't you search for publications that match the mutations we found in the Adenovirus genome."

"Why, what are you thinking?"

"Just a gut feeling…what was that?"

Jack and Amy were silent and then he heard it again. It sounded like glass breaking. Jack stepped out into the hallway and heard several voices from near the end of the building. At this time of night, Jack and Amy were generally pretty much alone. There may be the occasional scientist around, and a few security guards, but usually no one. Jack was already living on a short fuse so any noise out of the ordinary commanded his full attention. Jack heard heavy footsteps coming down the hallway towards the lab. This didn't sound good and he didn't think he wanted to find out who those feet belonged to.

"Amy," Jack whispered, "let's get out of here. I don't like this. It's too late for anyone to be here."

Amy didn't move from her seat, instead she swiveled around to listen more closely.

"C'mon, Amy, let's go."

"Dr. Pearson." Jack heard a shout from down the hallway. He looked at Amy. Why were they calling her? "Dr. Pearson, where are you?"

Amy stood up and looking at Jack, "They just want to talk to you about your work." She turned down the hallway in the direction of the oncoming footsteps. "Down here," she shouted.

Jack saw Aded round the corner of the lab bench and everything started to click into place. It wasn't Brandon that had alerted Intech that he was coming. That was why they hadn't stopped him at the guard booth when he used the fake pass. Aded didn't know how or when Jack was coming that night, only that he was going to be there and was going to try to get a sample of the virus. And then he looked at Amy and saw not the friend and lab mate he had come to trust and love like a sister, but rather a traitor. Someone who had been stabbing him in the back for a long time now, telling them God knows what; someone who had just set him up tonight.

Looking back he saw Aded striding down the hallway with two other men behind him. One of the men was Blondie, the man he had seen that night in Santa Rosa with Saresh. The other man wore small wire frame glasses and was quite a bit thinner than Aded or Blondie. Jack didn't think he had ever seen that man. Blondie had a large black gun with a long cylindrical barrel attached to the end. He pulled the gun up to aim at Jack.

Jack knew he had only a split second to make a decision. Did he trust Amy that Aded and the blonde man were only coming to ask him some questions at 11:00 o'clock at night at the lab, or should he try to make a run for it with a gun aimed at his back? There really wasn't much of a choice, he felt his life depended on this one run. Jack pushed Amy backwards and ran past her and through the computer room to the door on the far side. He heard Aded yelling something in a language that didn't sound English and he heard the scream from Amy, but Jack just kept running.

His heart was pounding so fast he was sure his vessel walls would not be able to handle the pressure, but he could no longer control the adrenaline that was shooting through his system. He ran past the benches in the lab on the far side of the computer room looking for anything that might be able to give him an edge or advantage over these men. He wasn't a fighter, and he had no gun. What was he going to do? He scanned the lab benches as he ran and saw Bunsen burners, lab books, small racks that might weigh a quarter pound and a few other items that would be useless as a weapon. It was ironic that a lab like this had developed what was probably the most lethal weapon known to man, but Jack could find nothing now that would help him to stop his attackers.

Jack swung around the wall and raced toward the break room. He spotted the full pot of coffee he had made earlier. He grabbed the pot, spilling some of the coffee on his left wrist and then slipped around the next corner. He could hear the heavy footsteps entering the break room and running past the coffee maker. Jack stepped out into the door opening just as Blondie was coming upon it. Blondie was surprised and quickly tried to pull his gun up but Jack threw the coffee in the direction of his face. Unfortunately, the coffee was slow to come out of the small opening; however, enough splashed onto Blondie's face to stop him in his tracks as he screamed in pain. Jack threw the pot at Blondie's head hoping to knock him out but it merely bounced off the side and crashed to the floor.

Jack turned to run again and felt a terrific blow to the side of his head. It felt as though someone had smacked him with a two by four board. His body continued forward but his head rocked backward. He stumbled forward in the darkness and then heard the rush of the ocean overcome him as he slumped to the ground. He tried to reach out and grab something to pull himself back up, but his arms wouldn't respond to the command. Beyond the roar of the ocean in his head he could hear a woman screaming and a loud, low voice yelling, and then he lost his fight with consciousness and slumped to the floor.

Chapter 32:

Jack slowly felt the world come back to him. He felt the throbbing pain on the side of his head and he could hear several muffled voices in the distance. He tried to move but could feel his arms pinned behind his back. It felt as though they were taped together. His legs were pinned together too. He tried to open his eyes and peak out and he saw he was inside Dr. O'Leary's office. In front him was Amy sitting in a chair watching his face intently. She too had her hands pinned behind her and her ankles were taped together with duct tape. Jack glared at the traitor in the chair in front him.

"I'm so sorry Jack," Amy whispered. "I'm so sorry. I didn't know they were going to do this. Jack, please."

Jack turned his head away from her. He didn't want to listen to anything she had to say. He had trusted her when he had told her he was going to steal a virus sample. Instead, she had put his life in danger and maybe Samantha's and Brandon's too. Oh God! What had he told her about Samantha? Had he told Amy about his cabin? Had he told her Brandon's name or enough about him so Aded would know who helped him? He had to escape and get back to Samantha.

"They told me they just wanted to know what you were working on. They said they were working on an anti-viral for the Adenovirus too. You don't understand my situation."

Brian J. Spencer

"Your situation," Jack hissed, "you put my life and my career in jeopardy. For what?"

"I needed the money. They already knew what you were working on. They just asked me to tell them how your research was going." Amy started to cry, "You don't understand. I needed the money. I needed the money for my Mom."

"If they only wanted to know about my research, why did you tell them I was going to Intech?"

"I just thought they should know that you were going to be poking around there. I figured they would stop you at the door if they knew you were coming."

"These are not people to be messing around with, Amy. Now both our lives are in danger."

"I'm so sorry Jack. I'll do anything to make it up to you. I promise."

"What do they want now? You must have told them I was here. What do they want from me now?"

"I don't know. Aded asked me to tell him when you were back in town. He said he wanted to work out a collaboration with you. He said he wanted to talk to you regarding your anti-viral protein."

"Yeah, well, somehow I don't think they really are too interested in any anti-viral research now. Intech isn't working on any anti-influenza or anti-adenovirus drugs, they're making a lethal biological agent. And with Bran..." Jack stopped himself. He couldn't remember how much he had told Amy before. He didn't want to give her or these guys any more information. "With what they are planning to do with that virus, they will want us to disappear. How much do they know we know?"

"I didn't tell them anything. They didn't ask. They just put me in here with you and have been out there the whole time talking on their cell phones."

Jack wondered why they wanted him alive. Or were they planning to take him somewhere else to kill him so there wouldn't be any trace here. Either way, he figured his life was on the line here. He had to think of something fast. It would help if he knew how much they knew and what they wanted to learn. Blondie poked his head into the office and called back to Aded who followed him in.

"So, Dr. Griffin, you're awake. I'm sorry to have to tie you up like this, but I can't have you running off again. Mike's going to have some real burns on his face because of you." Aded paused, apparently waiting for Jack to respond. "Let's start with what you know. You were quite clever at getting into Intech past our now former security guard. Tell me how you got in?"

Jack didn't answer. He wanted to give them answers, but he wasn't about to give away Brandon. Maybe they didn't know the two of them had worked together.

Maybe they didn't know Brandon and Samantha were out at his cabin.

"I see. You're not going to cooperate yet. How you got in is actually of no consequence to us. Let me tell you what I know you know. You have our virus. You know that it carries the ricin gene. And I have a hunch you know how deadly this virus can be. You also know we have a lot of it and have an automated process of producing it since you saw our BSL3 room. So now I have a few questions for you. Who have you told this to?"

"No one. I only learned of this tonight. I just checked on the mice and I only got back my sequence data tonight."

Aded looked at Amy who nodded her head. "I can't believe that a fine scientist such as yourself took five days to figure out what kind of virus he had in his hands. Where were you this weekend, and who did you talk to?"

So they didn't know that he had taken Samantha and Brandon out of town. They had probably been watching his apartment and the lab and had asked Amy to tell them when she saw him next. "I told you. I didn't tell anyone. I only got back to the lab on Sunday night. I was out camping in the desert with my dog. Waiting to..."

Aded reached out his foot and kicked Jack in the head. The searing pain raced across his scalp and he felt fresh blood drip down onto his neck. "I don't believe you! I want to know who you talked to! Who told you about us? Who got you into Intech?" Aded was shouting now. The calm demeanor he had portrayed for the past few minutes was fading quickly.

Jack felt the fear creep back into him. He had to think of something quick. He needed some time to try to think of a way out. Maybe if he could get them out of the room he could work his way over to Amy and have her help him get out of the tape. "I got an anonymous email telling me about Intech's plans. I can show you on the computer."

Aded rolled his eyes and looked back at Blondie. He motioned with his head to move out of the door where Jack could see Aded pull out his cell phone. This was going to be his chance. He wiggled his way over to Amy.

"I need you to work this tape off my arms. It's our only chance."

Jack sat up and pulled his hand up to Amy's arms so that they were back to back. Amy pried her fingers under the tape and pulled hard to try to tear to the tape off Jack's wrists but only pulled his arms back in the process. The tape was too tough to tear off. She next tried to dig her fingernails into the tape to rip it apart but slipped and dug one nail into Jack's wrist.

"Oww," Jack whispered.

Mike stepped back into the room. He reached over and ripped Jack away from the chair while calling back to Aded. But Aded didn't answer. Jack looked up and saw Aded slowly putting the phone back into his pocket while staring down the hall. Then he saw Aded sit down on the floor.

Mike pulled his gun out of his holster and put a shell into the chamber before hiding behind the door. Jack could see a man about 50 feet beyond Aded walking toward his side. The man was dressed in a black leather jacket and was wearing a black bandana over his head. He had on large leather boots that came up to the calf of his well worn blue jeans. As he approached Aded he grabbed him by the back of the coat and pushed him down to the ground. From around the corner, Snake showed his head. He had been the focus of Aded's attention.

Mike could see all this through the crack in the door and realized he was up against at least two very large men. He looked at Jack and put his finger to his lips as he pointed the gun at him. Then he reached over to the chair that Amy was in and pulled her closer to him while training the gun on her head. Obviously whatever favor Amy had held with Aded, Mike was not willing to convey on her.

The man who had pushed Aded to the ground was Chopper. Jack had talked to him many times at the shop, and he knew Chopper never went anywhere without Leroy who must be nearby. Jack looked directly at Snake without saying a word hoping he would get the signal that there was danger in the office. Snake appeared to understand and told Chopper, "Go look around. There might be more around here." Jack could clearly see Snake use his right hand, which was not visible to Mike, to point into the office.

Jack looked into the corner behind the door to signal to Snake and Chopper where Mike was hidden. Chopper turned to walk away but quickly turned and dived into the door with all his weight slamming into Mike. Snake jumped over Chopper who was rolling on the floor to pull a gun into a shooting position. The force of the door had dislodged the gun from Mike's hand, and he was just now reaching down to pick it back up. Snake pulled Amy's chair out of the way as Chopper slammed the door shut. He aimed his gun at Mike who froze at the sight of two guns on him.

"There was a third man, too," Jack told Snake.

"Yeah, Leroy's got him covered outside. He was standing beside the car. He should have him tied up and in the trunk."

"How did you know?"

"Chopper and Leroy were following the silver car you asked us to watch. They saw it come here and then watched these three guys get out and break the glass door to the building. They knew you and your lady friend were in here, and they figured there was trouble so they called me and I got up here as fast as I could. Looks like we didn't get here fast enough."

Snake had removed the tape from Amy's wrists and was now pulling the tape off Jack's wrists as well. Jack reached up and felt his head and saw the blood on his fingers. He might need a few stitches but it seemed his skull was still holding his brain in. Jack watched as Snake used the tape to bind Aded and Blondie. He wasn't gentle as he did it, slamming Aded's head into the ground and Blondie into the wall a few times as he moved him into the chair that Amy had been in.

"Look Jack, I don't know what you want to do with these guys, but if you're going to call the cops, we really can't be here. We don't exactly have the best reputation with them."

"No problem, Snake. Thanks for helping. By the way, if you were following these guys for the past few days, do you know where they were when they weren't following me?"

Chopper spoke up in his soft southern drawl, "They all weren't that easy to falla. One time we lost 'em goin out to the desert. They mostly went here or your place or someplace in the warehouse area downtown. I did see them come in here once this weekend when I knew you were out of town."

Jack looked at Amy who had turned to walk out of the room when she heard the news report of her meeting with her conspirator.

"You take care of yourself, Jack." Snake said. "Come on by the shop soon. I got a new seat for you to try out. I can put it on for you to try, and then I can take it off if ya don't like it."

Jack laughed. Snake was still trying to sell him parts. "Sure, Snake."

"And bring your lady friend here." Snake leaned into Jack, "I really like her. She's a good one."

"We'll see." Jack didn't want to get into that with Snake right now.

Snake turned and was followed down the hall by Chopper.

Chapter 33:

Jack had to think of what to do next. He could call the police for these two guys and the third out in the parking lot, but he still needed to get a hold of someone to warn about Intech's virus. Jack walked back to the office to use the professor's phone to call the police for Aded and Mike. He then called the professor to explain the situation.

Jack returned to the hallway to find Amy leaning against the far wall. "Do you know how to use a gun like this?"

"No, I…no, I've never held a gun before."

"Well, here's your crash course. Here is the safety. You have to push this in if you need to shoot. There's a shell in the chamber now and the gun is cocked. All you have to do is flip the safety off and pull the trigger. I want you to watch these guys while I destroy the mice and cells I have. I don't want this virus getting out of here."

"Wait Jack. I wanted to tell you what I found out about the virus. I searched for published articles that described the mutations you found in the Adenovirus."

"Yeah?" Jack really didn't think this was going to go anywhere. Amy was trying to stall for time so he wouldn't leave her with Aded and Mike.

"The first two sets of mutations were published in an article in 1979 from here at the Hoffman Institute. The mutations increase the replication rate of the virus. The paper is on the printer now."

The news stunned Jack. Not only had these people put a more lethal form of the ricin toxin in the virus, but they had also used a modification of the virus to increase the infectivity. These mutations would result in the virus spreading faster through the population resulting in an impossible containment scenario. What were these people trying to accomplish?

Jack pulled the paper off the printer and scanned over it. He didn't recognize the lead author's name Sa Reshdi, but apparently he had been at the Hoffman Institute. Jack paged through to the end and saw that this work had also been funded by DARPA. Jack started to toss the paper aside to take care of the mice when he glanced at the authors again and it hit him. Sa Reshdi. Could it be the same Saresh from Santa Rosa. Jack sat down at the computer and did a search for other papers published by Sa Reshdi, but couldn't find anything more after the 1979 paper.

DARPA had funded Dr. O'Leary's research on the ricin toxin and Sa Reshdi's research on the Adenovirus in the 70's. Both had published papers describing mutations that would lead to increases in toxicity, and then Sa Reshdi appeared to leave science. Now it would appear he had turned up in Santa Rosa connected with a company that had an Adenovirus with his published mutation containing the ricin toxin with Dr. O'Leary's published mutations. Something didn't add up. Had DARPA been funding research at the Hoffman Institute to develop a biological warfare program in the 70's? That would have been ten years after President Nixon signed a treaty banning all research on biological weapons in the United States. And now, Jack and Amy were funded by DARPA looking at ways to stop the spread of the virus and block the action of the toxin. Of course, Dr. O'Leary had asked Amy to apply Jack's transporter protein technology to her anti-toxin project. They would need to deliver the anti-toxin to the nose, throat and lungs in an attempt to counteract the effects of this virus. But did Dr. O'Leary know about Intech? Did the government know? That might explain why Dr. O'Leary had come down on Amy so hard about getting some results soon.

Jack knew what he had to do. First thing was to destroy the animals and the cells for the safety of everyone at the institute. He still had a sample of the virus stored in the freezer. Autoclaving everything that had contacted the virus would destroy the danger. Jack put on a smock, hair net, face mask and finally two pairs of gloves before heading into the room with the cells. He pulled the dishes with the virus samples out and placed them into an autoclave bag. Next he placed this bag into another autoclave bag to prevent any leakage. Then he sprayed down the whole inside of the incubator and all surfaces he touched with 70% ethanol. Next he went down to the animal room to recover the mice. Whenever mice died under mysterious circumstances, the animal care technicians put the bodies in the freezer to allow the investigator to examine them. Jack pulled the mice out and double bagged them too. Finally he placed all the waste into the autoclave and started the machine.

Jack returned to the professor's office to find Amy sitting in the chair in front of the professor's desk.

"Did the police come already?"

"Yeah, they took Aded and Mike. They're looking around the institute to see if there are others here too, and then they will be back to talk to us. Jack, I ah…"

Jack walked out of the office without allowing Amy to finish the sentence. He would wait for the professor in the hall. He didn't want to confront Amy right now.

"Jack, Amy what's going on here?" The professor asked as he walked into his office. He was wearing sweat pants, a t-shirt and a long coat that he was using in an attempt to cover himself.

"Hi John," Jack started. "What can you tell me about your research with DARPA in the 70's and the work of Sa Reshdi during the same time?"

Dr. O'Leary looked startled as he looked from Jack to Amy and back to Jack before looking down at his feet. Closing his office door behind him, he slowly crossed to his chair behind his desk and motioned for Jack and Amy to take a seat.

"I guess you startled me there. I didn't realize you knew that much. I don't know what you really want to know."

"A company named Intech here in town," Jack said, "has developed a virus with mutations that were described in a paper by Sa Reshdi. These viruses are carrying the ricin toxin containing mutations described in a paper by you. Both projects were performed in the same lab and both were funded by DARPA. What's going on here, John?"

"You have the virus? Where is it?" Professor O'Leary's face went white.

"I have the virus. Once I found out what it was, I sterilized everything associated with it by autoclaving. I still have a sample stored at $-135°C$. C'mon John, what's going on? We need to do something, but we need to know what is going on. People's lives are in danger here."

"Okay, when I first started out here at the Hoffman Institute in 1975, I was a post-doc, like yourselves, working for Professor Jeffrey Tollner. Dr. Tollner was young to have his own lab at the Hoffman Institute, but he had published several groundbreaking papers on the study of viruses as biological agents. This was a time when most people were focused on smallpox or the newly discovered hemorrhagic fever viruses. Jeffery was performing research studying the use of normally innocuous viruses such as Adenovirus and Influenza as bio-warfare agents.

"Funded from a grant from DARPA, I started in the lab studying ricin toxin as a biological agent. Most of the lab was funded by DARPA, but we didn't ask any questions. We thought if we found out how these viruses and toxins could be used as weapons, we could develop therapies or treatments for them.

"Sa Reshdi came to the institute after I had been here about a year. He was Pakistani but had received his Ph.D. at MIT. He was brilliant. He studied the Adenovirus, applying select mutations to the virus genome and analyzing them for increased virulence. He published a paper on the virus, but then he moved on to a new project. He told us all he was working on a new project for DARPA, but we didn't believe him. He was a genius and we all thought a little crazy because of it."

The professor turned to look at a photograph on the cabinet behind his desk. "We called him Spanky, although I don't think he liked that too much. He and I worked together a lot. I remember after he published his paper on the virus mutations, he wanted to celebrate and so just the two of us went out for dinner and drinks.

"It was shortly after that that he became more withdrawn. He was never one to join the softball team or play volleyball with us, but after that paper was published, he started withdrawing from his girlfriend, Carolina and me. He only wanted to work in the lab. Sometimes he would be there for two or three days in a row. None of us knew what he was working on anymore, although we all assumed it was another project related to his original virus work."

The professor turned back to face Amy and Jack. "And then suddenly he disappeared. Men from the military came looking for him, ostensibly because he was funded by DARPA. By now, many of us were starting to question the projects he had worked on, but he kept his lab books in a code that no one could decipher. We all moved on from the lab over the years, but then about a year after I left, Dr. Tollner was murdered in his laboratory late one night. His lab was ransacked and set on fire. No one had seen a thing, but the police had pulled several finger prints of Sa Reshdi's. They suspected him, but again, no one saw him."

"So do you know what happened to him?" Jack asked.

"I never heard from Sa again, but DARPA came to me about five years ago and asked me to work on a special project. As you know, most grants are competitive with a notice going out to ask investigators to submit proposals for a project, but this time, they came to me and told me I would get the money without writing a proposal. I guess I knew what they wanted the day General Osterman came to see me. They told me a lot more of what had been going on in those days in Dr. Tollner's lab. We hadn't been working to develop treatments or therapies against potential biological warfare agents. We had been working to develop those agents for our military. And we did. Sa Reshdi had been working on a special project to develop this ultimate weapon. General Osterman assumed Sa Reshdi had developed that agent, and then had taken it with him. At first we assumed he had taken it because he found out what he was developing and why he was developing it. The CIA had begun to search his background more intently after his prints showed up in Dr. Tollner's lab following the fire. In the days before he disappeared, it seems Sa Reshdi was meeting with members of a Pakistani militant group.

"Now, they were convinced he had taken the virus. They came to me and funded my projects to develop therapies against the toxin and the virus. That's what you two have been working on. It's ironic, I spent years as a post-doc developing one of the world's most deadly viruses, and now I am spending a lifetime developing an antidote for it. Hopefully before it's too late.

"So, now you tell me what you know. Wait, let me get General Osterman in here."

"I think that would be a good idea." Jack said. "We need to stop this company, and I don't think the local police are equipped to handle this."

Chapter 34:

It was after 2:00 AM when the General strode down the hall followed by two military police carrying holstered side arms and scanning the lab space for threats. The local police had put guards at the outside doors in case anyone else tried to show up that night, although no one really thought that would happen.

"Gentleman, let's use the office." The general led everyone into Dr. O'Leary's office leaving the two MPs outside the door.

"General Osterman, these are two post-docs of mine, Jack Griffin and…"

"Amy Pearson. I know who these two are. Let's get down to business. You said on the phone that you had information about Project Wildfire."

"General Osterman," Jack interrupted, "I'll make this short sir. I took a sample of a virus from a company called Intech, which is located downtown in the warehouse district. I obtained the sample on Friday night, and since then I have determined that the virus is an Adenovirus containing specific mutations known to increase the virulence of the virus. These mutations were described in a paper written by the professor in 1978. The virus also contains the ricin toxin gene with mutations described in a paper written by Sa Reshdi in 1979."

"How did you get a hold of this virus?"

"I uhh, I, well. Isn't the point that we know it exists?" Jack didn't want to pinpoint Brandon in this.

"Right," the general responded questioningly. "What else do you know?"

Jack needed to tell them about the wind study and their plans to release the virus. He couldn't think of any other way but to just tell them about Brandon. He could at least leave out Brandon's help in his breaking into Intech. "The reason I got involved in Intech is because a friend of mine came to me for help. He has been working for them, and he thought he might be in over his head. He started with Intech taking wind measurements from about 100 yards off shore. I think Intech may be planning to release the virus into the wind to distribute it over the whole city." Jack stopped and looked at the professor and then at the general who appeared lost in his thoughts.

"I need you to tell me everything you know about Intech. I'll assemble a team to take control of the building."

"One more thing, sir. I think I know where Sa Reshdi is. He is going by the name Saresh and he has an office in Santa Rosa."

The general put the phone back down and turned to look at Jack. "This is indeed good news. I want you to come along with us to take down Intech, and I'll send a team to Santa Rosa to bring in Sa Reshdi or Saresh or whatever name he is going by now."

Jack explained the layout of the lower floor of Intech to the general who took meticulous notes. Jack was surprised by the level of knowledge the general had regarding laboratories. He described the location of the BSL3 facility and the freezer rooms. Those would be the two locations in the building that the general would want to secure first.

As Jack was finishing his description of the building, an MP guard came into the room and whispered into the general's ear. Jack saw General Osterman sit straight up and watched as the color drained from his face.

"It appears we have a problem. The advance group I sent out to watch Intech has reported that the building is on fire. The whole warehouse district is on fire now. There is some evidence from people in the area that a large explosion occurred in the vicinity of the Intech building. Fire crews are working on it, but they don't expect that we will get in there before first light."

"They destroyed the evidence," Jack said. "They must know we have Aded."

"Is there anyplace else you can think they might be?"

"Only Santa Rosa where I told you Saresh was."

"Yes, well we have a group heading up there now, but personally I don't think we'll find him now. Is there any danger to the public from this fire?" The general asked Dr. O'Leary.

"No, the heat from the fire will destroy the virus and any toxin that might be around."

"Is there any possibility they could take the virus with them?"

"I would have to say yes. They could store the virus at −20°C for a short period of time. And that could be done in any household freezer."

"How long are we talking about here?"

"The virus will slowly begin to degrade at −20°C, but you would probably only see a 50% loss in a week."

"I see," said the general. "If Jack is right and their plan is to release the virus into the air to spread over the city, we'll need to patrol the airspace over the city and in particular over the ocean." The general turned back to Jack, "Is there anyone else besides this friend of yours involved in Intech?"

Jack looked at Amy. He was deeply hurt that his friend would betray him like she did. He was sure he would never be able to fully trust her again, but he couldn't turn her in for what she did. He believed her when she said she had to help her mother. He had heard her talk many times about how poor her mother was and the fact that her father wouldn't support her. Jack looked back at the general, "No sir. No one else."

"Okay, I'll want to talk to this friend of yours."

"His name is Brandon Singer. He is staying at my cabin outside of town for fear of Aded and Saresh. I'll get to him first thing in the morning and bring him to you."

Chapter 35:

It had been a long night already. The discovery of the virus and the shockingly fast death of the lab mice had scared Jack. He had a splitting headache from the hit from Aded; and now his conversation with the general. The sun was just starting to come up over the mountains to the east and was now lighting the lab through the large glass windows. Jack wanted nothing more than to head back to his home and sleep for two days. He was concerned by the destruction of Intech and the army's lost chance to gather any evidence and seize the virus. He was also concerned by the potential escape of Saresh. He was sure that this plot would be repeated if these people weren't caught now. On the other hand, with the military now searching for Saresh and patrolling the skies for the release of the virus, hopefully this would all be over soon.

Suddenly he felt a renewed urgency to work on his project of developing a good anti-viral drug for the Adenovirus. He was sure that with the help of Amy, the two of them could develop at least a preliminary drug against this toxin-carrying virus within a few months. He just hoped he had that much time.

Jack left the office followed closely by Amy as he headed toward the break room for a cup of coffee. It was then that he remembered that he had smashed the pot trying to stop Blondie.

"Jack, please." Jack turned to face Amy for the first time since learning of her betrayal. He felt sadness in his heart at what he saw. Her hair was a mess and her eyes were puffy from crying. She had a black eye forming from what Jack could guess was a hit while he was out cold. He really felt sorry for her. He wanted to trust her again. He wanted to face his friend and lab mate again. He wanted to work with her as he had in the past. Maybe he could forgive her given some time, but now was just too soon.

"Thank you for not telling the general about me. I thought I knew what I wanted to say to you, but now..." Amy started to cry again, but Jack made no move toward her. "I never thought they would hurt you Jack. I didn't think I was putting you in jeopardy. I just really needed the money, and I thought the information they asked for was harmless. I know I was giving them information about your research, but you've got to believe me, I never told them about your big breakthroughs. I never told them about the transporter protein idea of yours. I wanted you to publish that. I figured I could string them along for a bit and then tell them after you had submitted a paper for publication but before it was printed. Jack..." Amy turned away and Jack could tell she was really crying now.

Jack felt his emotions overwhelming him. He hated seeing his friend in pain this way. "Amy, it's okay..."

"No, it's not okay. I lied to you. I betrayed your trust. I loved you, Jack. I love you...and I did this to you."

She turned back to Jack and allowed him to pull her close to his chest. "I love you, Jack. But now I don't even deserve you. I'll do anything, anything for you to make up for this. Just so we can be friends again."

Jack allowed her to cry into his chest as he thought of the ramifications of what she had just said. He should have known she had feelings like this for him, but on the other hand it was hard to reconcile her recent actions with her proclamation of love for him. They say time heals all wounds. Jack was willing to allow himself a chance to forgive her, but he knew he would need time.

"There is something you can do for me now. I want to head back out to my cabin and bring Brandon back here so the general can question him, but I want to stay out there with Samantha. I just need a short break from all this. I'm tired, my body is beat up, and my head is beat up. I need a couple of days to rest and then I'll be back here to work. I need you to drive the Jeep out and bring Brandon back to town. I'll ride my bike out and stay there."

Amy looked down. "I'd love to help, Jack. Anything you want." Amy chuckled, "I was just hoping I could ride with you on your bike again."

"Tell you what. You drive the Jeep out there, and I'll give you a ride before you head back to town."

Amy seemed to brighten up some. "Anything, anything. I just don't want to lose our friendship."

"C'mon, we'll get some steaks and beer before we head out and then we'll all have a quick dinner before you and Brandon come back here."

Jack gathered up his stuff and explained to the general that Brandon would be in touch with him by Friday. Jack led Amy out to the parking lot and to his Jeep. He found himself subconsciously looking for the 300ZX. He couldn't believe this was over and he could finally spend some time with Samantha without having to worry about what he was going to do next. And she would finally be able to relax with him without worrying about her brother.

Jack drove Amy to his apartment and gave her the Jeep with a cooler of steaks and beers as well as some asparagus that he planned to try grilling. He then went around to the back of his apartment and pulled his bike out from the parking garage. It felt good to sit back on the Harley, and as he turned the key and kicked the starter, he felt as much as he heard the tha-thump, tha-thump of the big V-twin engine. Sitting on the bike waiting for it warm up, he realized that he really didn't want to try a new seat on the bike. In fact he didn't want to try anything new for a while. He wanted to cruise for a while and see what developed with Samantha.

Jack put the bike in gear and slowly pulled around front, honking to Amy. He rode down the street and headed east to his cabin, with Amy following in the Jeep.

Chapter 36:

The sun was full overhead by the time he rode down the road his cabin was on and past the familiar sign '15 Miles to Jack's Shack'. He had barely pulled into his driveway when Ace came running out to greet him.

"Hey buddy. It's good to see you too."

Jack looked at the porch around his house. It hardly looked familiar with the dark blue paint on the floor and columns with a bright white trim around the railing. Samantha stepped out onto the porch wearing a long white summer dress that barely clung to her shoulders. She was holding two glasses of lemonade as though she had been waiting for him to arrive for a long time. She was exactly the sight Jack had wanted to see today.

"Hey, stranger. I thought I heard your bike so I grabbed you a glass of lemonade."

"Ohh, you are a sight for sore eyes," Jack said as he swept onto the porch and grabbed Samantha around the waist. "And so is that lemonade. It's damn hot today."

"Santa Ana."

"What's that?"

"Santa Ana winds. In the winter sometimes, the winds shift from blowing off the ocean to blowing from the desert and it really heats up the land. That's why it's so hot today. Hey what happened to your head?"

"Let's just say that we don't have to worry about Aded anymore. Brandon can go home safe. The police have Aded and are looking for Saresh now. I'll tell you about it over dinner."

Amy pulled into the driveway with the Jeep causing a swirl of dust to blow past Jack and Samantha. She got out and walked around to the back of the truck to gather the food.

"Samantha, this is Amy, my co-worker."

Samantha smiled broadly, "I've heard so much about you Amy." She grasped Amy's free hand, "Jack says you're the one person he would trust with his life."

Amy looked a bit shocked and then lowered her eyes to the ground. Samantha looked at Jack and back at Amy before noticing the 'don't ask' look on Jack's face.

"Amy's going to stay for dinner. We brought some steaks and beer, and then she's going to take Brandon back to the city. Some people want to ask him a few questions about his involvement with Intech. Don't worry; he's not going to get into more trouble. They just need all the information they can get to find Saresh and the rest of the people from Intech." Jack looked past Samantha into the house. "Where is he?"

"He's around back with Bill. The two of them have been painting the barn since you've been gone. Bill comes over every day and the two of them take their time painting."

"I'll, ahh, put these inside for later." Amy stepped inside leaving Jack and Samantha alone on the front porch.

"I like what you did here."

"I painted the porch during the days. There isn't much to do here. No TV, no radio, not even any books. I found some paper, and so I've been writing in the evening. I was so worried about you though, I couldn't concentrate. I'm so glad you're back." Samantha reached up and touched the sore on Jack's head. She leaned in and kissed the bruise forming next to his eye and then kissed him on the lips. She kissed him slow and deep to let him know how much he had been missed.

"I was thinking we could spend a few days here, just the two of us. We could get back by Sunday night."

"I think that's a great idea. Somehow, I think with you here, I won't be too concerned about the lack of entertainment," Samantha said with a smile. She then turned to head around the house toward the garage.

Amy was already standing in the barn doorway talking to Brandon and Bill when Jack arrived. The three of them were drinking beer from the bottle huddled in the shade to avoid the already oppressively hot sun.

"So I hear you've been busy causing trouble, Bill."

"Ahh, umm, Ahm sorry, Jack. I was jus' helpin' a bit."

Jack grinned at the pained expression on Bill's face. "It's okay Bill. The barn needed a good painting anyway. So Brandon, I've got good news. You can go home again. You don't have to worry about Intech or Aded anymore."

"What about Saresh?"

"There are a lot of people looking for him now. I think it's safe. There's a General Osterman who wants to speak with you about Intech. He's going to be in charge of finding Saresh and picking up the pieces of Intech."

"All this over the flu?"

"Well, actually it wasn't the flu virus."

Amy stepped in, "What Jack found at Intech was probably the most dangerous virus ever. Imagine taking the ability of the cold virus to spread quickly and easily across populations and adding to it a toxin that has been known for centuries to kill humans with the speed and efficiency of a guillotine. Intech had created and grown the most lethal bioterrorism weapon."

"How could they do that? I mean, is that even possible?"

"Possible? Definitely. It's incredibly easy to do. The sequences for the virus and the toxin have long been published, and in the case of the virus that Intech had developed, descriptions for increasing the toxicity of the virus and the toxin have been published."

"I don't understand. You mean even I could find this stuff?"

"Yeah," Jack spoke up. "It's just that easy. We publish everything we discover in science, and the journals are available at the library."

"Why would you want to publish this stuff? I mean what good could come from putting out a paper that describes how to increase the toxicity of a virus?"

"Sometimes the benefits aren't easily apparent to people not in the field. If I wanted to use the virus for a beneficial approach to cure a disease, I would want to know what changes to avoid. Or if I were trying to develop a cure for the virus, I could use that information to identify a gene to target. So this information is necessary for people in the field."

"I just can't believe this information is right out there. So they got the virus now?"

"Intech burned down and that would have destroyed the virus, unless they moved it out. Don't worry though, they wouldn't have been able to move out that much and keep it cold. The military will be hunting them down now to find out if Saresh tried to set up a new lab."

"I can't believe I was helping them." Brandon backed up and fell into the couch.

Samantha walked over and sat down next to him. "Brandon, you didn't know. You couldn't have known what they were planning."

"But maybe I should have. I mean, c'mon, giving people the flu so they could test some drug. You wouldn't have fallen for that."

"Brandon, c'mon," Jack said. "You're not supposed to know the difference between a harmless flu virus and a genetically engineered weapon. No one can tell that. I study this stuff every day and I would have believed them."

"But you didn't believe them."

"I believed you when you told me the story about the flu. It seemed like a logical explanation. I just wanted to see what flu virus they had. You had the guts to step up and help me. You put your life on the line getting me into Intech. I know that. Maybe you got in over your head, but you stepped up when it counted."

"Brandon," Samantha spoke softly, "I'm proud of you. You did the right thing here. I'm sure Mom and Dad would be proud of you too."

Jack looked down and saw Ace lying at his feet with his Frisbee in his mouth. "C'mon buddy, let's go out and play."

Amy and Bill followed him outside. "Didja really save us from some killa virus, Jack?"

"He did. He's a real hero. He even saved my life last night."

Jack blushed, "I don't know about hero. I think I just did what anyone would have done in that situation, Amy. Besides, I know viruses, so it really wasn't much work for me to find out about this one."

"He's being way too modest, Bill."

Chapter 37:

Jack hoped to deflect the conversation, so after a few throws of the Frisbee, he said to Amy, "How about that ride now?"

"Okay, but can we skip the helmets today? It's just so hot."

"Yeah sure. It really is hot today. Samantha called it the Santa Ana winds. She said wind gusts will reach 30 or 40 miles per hour, but it shouldn't affect us too much on the bike. Just hold on tight and let me correct for the wind. I'll stick to the country roads around here so we can skip the helmets."

Jack threw his leg over the seat of his bike, pulled the choke lever and reached down to turn the key. Jumping up off the seat he landed with all his weight directly on the kick-start lever on the right and listened as the engine roared to life. He eased back on the choke and heard the familiar tha-thump, tha-thump of the motor. He motioned for Amy to hop onto the back of the seat and then handed back a pair of glasses that would protect her eyes from the flying dust in the air. Jack eased back on the throttle and roared down the driveway turning right onto the country road he shared with Bill.

The wind was a bit rough, gusting around the bike making it feel like a large hand was flicking the bike from right to left causing Jack to perform constant corrections. He rode on down the long straight road to the east past Bill's place and toward a small depression in the earth. Here a small creek had run north-south a long time ago, but it had been dried out by the over-farming of the area. In fact, the government had revoked the irrigation rights of the farmers in this area almost ten years ago to accommodate the increasing growth of populations in the coastal cities. A lot of the farmers just left, leaving a lot of dry, barren land that would take decades to recover. Jack's land had once been used to grow citrus trees much like Bill's land, but now, only cactus and scrub brush could survive. The trees had long ago given up their roots in this land.

The road turned to the south toward the foot hills that ran parallel to the coast, but Jack had no intention of riding that far. He was tired from the long night and the long week that he had endured. He couldn't wait for that steak dinner and then a nice quiet night in front of the fire with only Samantha and Ace. That would ease a lot of the stress from the past few days.

He thought of Amy on the back of his motorcycle and felt her arms around his waist as she hugged him close on the ride. Although he had suspected she wanted them to start dating, he was still surprised to hear her say she loved him. He would really like to forgive her and forget what she did to him. He missed his friend and co-worker from just a few days ago. He had always thought she was the one person in science he could fully trust with even his most unconventional ideas. One day he hoped to have that relationship with her again, because he couldn't imagine how hard his job in the lab would be without her support. In the meantime, though, he would have to watch her more closely and be a bit more careful of what he said around her.

Jack turned the bike around on a slightly wider section of the road and headed back to the turn in the road at the creek. As he approached and slowed for the turn he felt a tug on his arm and saw out of the corner of his eye Amy's hand pointing toward the sky. Following her finger he saw a crop dusting plane flying close to the Earth. Jack turned onto the road he lived on and pulled over to the side to watch the plane more closely.

"Why is that plane flying so low?"

"It's a crop-duster. But I don't know why it would be out here. There's nothing growing out here now."

"I would think it would be dangerous to fly so low with the wind blowing that…"

Jack turned in his seat and saw Amy's face was as white as a sheet. Her lips trembled and tears formed at her eyes. Jack was looking into the face of terror.

"Jack," she whispered, "oh, my God! Jack, what if Intech never intended to release the virus into the air over the ocean."

"Well, I guess it's possible, but why would they have had Brandon make all these wind… "And then Jack understood Amy's terror. Intech used Brandon to measure wind speed and direction. They weren't measuring the wind for releasing the virus into the easterly breeze off the ocean. They wanted to release the virus into the westerly Santa Ana winds. The strong winds would allow them to release the virus a greater distance from the city. The dry air wouldn't destroy the Adenovirus, but it would have destroyed the flu virus, which needs a humid or wet environment to survive. The military was looking in the wrong direction and even now they were two hours from here. Jack watched as the plane lined up to the right of the road and swept down to land on the far side of the new barn.

Bill had mentioned that the area around the barn had been strung with barbed wire and armed guards were stationed at the entrance. There was no farming going on there; that was Intech's satellite location. No wonder they torched the warehouse building; they had probably intended to move everything out here.

How would they have moved the virus out here? Maybe it wasn't too late. Maybe they were only setting up a new lab. Jack knew he had to get in and find out fast. Overnight the winds were supposed to calm down and switch back to a normal easterly flow. If they had the virus and were planning to release it now, he would have to get General Osterman out here now. Jack started the bike and rode as fast as he could back to his cabin. The wind whipped his hair against his ears and the pants legs against his legs causing his skin to go numb. The speedometer needle crept up over the 90 MPH mark as Jack tucked his head down and into the oncoming blast of air.

He could feel Amy gripping his waist even tighter. She had wrapped her arms all the way around his waist partly due to fear of the speed and partly due to fear of what Intech may have planned. The release of small doses of a virus like that would infect thousands of people by morning. Those infected with such a small dose would probably fail to show symptoms for a day or two allowing them to spread the virus to others. Those boarding flights to other parts of the country would spread the virus quickly throughout the whole nation and ultimately throughout the world. Death would come quickly to those infected, probably within six to twelve hours of the first symptoms. Unfortunately, because this was a genetically engineered virus, no one would have natural immunity to the infection. Jack felt like a gladiator who had just learned he was to fight not one but ten lions in the ring.

Chapter 38:

Jack slid the rear tire around under the bike as he turned into the driveway of his cabin and spun the tire as he accelerated toward the garage behind his house. He flipped down the kickstand as he leaped from the bike.

"Brandon! Brandon!" Jack yelled. "Brandon!"

Brandon came running out of the barn, his hands covered in grease and oil. He was followed by Samantha and a slow lumbering Bill. "What? What's wrong?"

"Is it possible you were measuring the Santa Ana wind flows and not the normal breezes off the ocean?"

"What are you talking about?"

"Intech!" Jack screamed, "Did they ask you to measure the winds from the east or the west?"

"They told me to measure the winds everyday and to record the speed and direction. Most of the days it was from the west, but there were a few days last month when the wind blew from the east. Why?"

Jack saw the comprehension in Samantha's eyes. "Oh my God! No!"

Brian J. Spencer

Brandon spoke again, "They did seem more excited the day the Santa Ana winds blew. They told me I had to be out those two days. In fact I had to go out twice each day. In the morning and evening. Why?"

"I don't think they were at all interested in an ocean breeze. I think they planned all along to use the stronger, drier Santa Ana winds from the east. I think they have the virus here, down the road at that new barn."

"We need to call this general guy now. Someone needs to find out about this," Samantha pleaded, seeing the look in Jack face.

"I don't think there's time. Brandon, run to Bill's place as fast as you can and call that number I gave you. It's General Osterman's cell phone number. Tell him where we are and what I just said. Tell him to get people up here now. I need to get over there to see if they are planning anything for this afternoon. The winds are supposed to die down tonight and then switch directions. If they are planning something for today, they would want to do it in the next few hours."

"Jack, no please," Samantha pleaded with him. She ran to him and put her arms around his shoulders. "Please, let the general handle this."

"Samantha," Jack pulled her to the side away from everyone. "I don't think they can get here fast enough. You said that sometimes a person needs to step up. This is my time. I can't sit here and let them release the virus on everyone. I saw it. I know what it can do. I can't let them get away with this. Besides, this may be nothing and I'll just be right back," Jack lied to her.

"Brandon, get going. I'm going over to the barn now."

"No!" Amy shouted startling Jack. "I'm going with you. You need someone to watch your back, and I owe you."

Jack hesitated. He didn't know if he should really trust Amy yet. "I don't think so."

"I don't care what you think, Jack. I am going with you. You don't get to decide this."

He looked into her eyes trying to determine if he could rely on her. It would be nice to have someone else with him; another pair of eyes to watch his back. And besides, maybe she wanted to get back at Saresh for using her. Maybe she really was sorry and he could trust her. "Okay, wait here one minute."

Jack walked into the house followed closely by Samantha. He headed for his room and reached under the bed to retrieve the wooden case housing the Colt 45.

"Jack, please don't do this. I need you here. I need you," Samantha began to sob.

Jack put his leather jacket on and slipped the gun into the side pocket and a small pair of binoculars into his inside pocket. "Samantha, I'll be back. Don't worry," Jack said with a false confidence.

"I love you," Samantha said softly.

Jack stopped in his tracks and turned around to look at Samantha. That was the second time he had heard that phrase today, but this was the one woman he wanted to hear it from. Jack stepped closer and kissed her lightly on the lips. "Hold that thought, I'll be back later."

Chapter 39:

Jack turned the bike around, and Amy hopped on the back of the seat. Samantha was still inside the house, and Brandon was already on his way to Bill's house. Bill was still standing in the doorway of the garage apparently waiting for the world to end.

Jack eased onto the throttle this time and pulled the bike out onto the road. The two of them road past the steel fence on their right and noticed the ring of barbed wire along the top. The driveway to the compound was secured with steel gates, and standing outside were two very large men guarding the entrance. Jack couldn't see any guns, but he felt sure they had weapons hidden under their heavy jackets, which were odd apparel in this hot weather. Jack rode on past the compound and looked at the new steel shed located about 200 yards from the road and which ran parallel to the road. After passing the barn, they could see the landing strip on the far side of the structure with the same crop dusting plane they had seen flying overhead earlier. Jack rode on about a half mile before pulling his bike off onto a small dirt path. He hid the motorcycle behind a small scrub bush and pulled another dry, dead bush against the bike so the wind would hold it against the front wheel and camouflage the machine.

"Now what?" Amy asked

"We need to get closer so we can see what's going on."

Jack headed out on foot toward the rear corner of the fence assuming his enemies would be watching the road more closely than the rear of the compound. Although the sun had arced over the center of the sky and was starting to head back to the west, it was still very hot as Jack and Amy walked across the desert floor. Jack pulled the binoculars out as he approached the fence. Using a small cluster of agaves cactus as cover, he crouched down and looked into the compound. From this angle, he could clearly see the plane and the open side door of the barn, and what he saw drove the fear of God right into him.

Parked just inside the barn was a large refrigerated truck. Along the inside wall of the barn he could see several large freezers similar to the −80° Celsius freezers he saw at Intech. Several people wearing full space suit-like protective outfits were loading large cylinders onto the wings of the plane. The stainless steel cylinders were approximately three feet long and a foot and a half wide with one tapered end and one square end for standing upright. They looked an awful lot like bombs, but Jack feared they would be much more deadly than a mere explosive device.

Jack handed the binoculars to Amy, "Here look."

After a minute Amy spoke barely in a whisper, "They're going to do it now. We need to do something."

"Okay, we need to have a plan. I only have one handgun and I wouldn't exactly say I am a crack shot. This isn't TV or a movie. I don't think we can shoot our way in there and stop them."

Jack looked at the crew around the plane again through the binoculars. He saw two men wearing the protective space suits and a third that was taking his off beside the door to an office. Inside the office he could see Saresh and another man sitting in chairs. An air conditioner was mounted on the wall facing Jack and their only view into the barn was through a small window in the office door.

"I need to get in there to see if I can destroy the virus."

"How are you going to do that? There are three people working right there with the virus."

"Actually two. One just went into the office. I guess I'll figure out what to do once I get there. Watch me with the binoculars. When I get into the barn, get their attention some how so that I can find some way to destroy the virus. If I can set the barn on fire, I can destroy everything at once."

"How am I going to get their attention?"

"I don't know. Just do whatever you can."

Jack looked up at the fence and the barbed wire running along the top of it. He took off his leather jacket to throw over the barbed wire. "This was my father's jacket." Jack ran up to the fence with the gun tucked into his pants and threw the jacket up onto the barbed wire. Taking a few steps back, he jumped onto the fence reaching his arms over his jacket. He quickly flipped his legs over and dropped down inside the compound.

Jack looked around, but it didn't seem to him that anyone had noticed. Crouching low to the ground and using various cacti and boulders as cover, Jack made his way to the plane where he sat down on the backside of the wheel. The plane was a simple one with a single seat inside a closed cockpit. The one propeller was mounted in front of the aircraft and the tail dragged on the ground.

They had loaded one of the large cylinders onto this side of the plane, and now Jack could see there were mounting points for two cylinders under each wing. A coiled hose ran from the top of the cylinder to the spray nozzles located along the length of the underside of the wing. Normally these cylinders would hold some insecticide for spraying crops.

Jack stood and inspected the cylinder that was already attached to the wing. He couldn't think of any way to destroy the virus inside this cylinder, but he could uncouple the hose. Jack pulled back on the clamp and popped the hose off the nipple on the cylinder. The other three cylinders still must have been inside the barn.

Jack looked back to Amy but he couldn't see her now. She was well hidden behind a cactus cluster. He waved back in her direction more to give himself some confidence in his mission. He suddenly realized how much he trusted her now. He hoped he was right on this.

Jack stood up and looked through the windows of the plane and into the barn. He could see the two men toward the far wall of the barn with one cylinder perched on a wooden dolly to allow them to wheel it out to the plane. Jack ran around the corner of the barn just outside of the view of the people inside. From this point he could see the two men at the front gate; however, they had their backs to him. Jack eased around the corner and saw the protective suit, which belonged to the third man, hanging on a peg near the office door. Jack pulled the suit off the peg and pulled it back around the corner with him. He slipped into the suit and put the full-face mask over his head. In less than a minute, it was already becoming unbearably hot.

Jack stepped inside the barn and got his first real good look inside the structure. He saw the freezers lined up along the back wall; however, two of the freezers were shaped more like brew kettles. He could see frost had formed on the outside of the kettles indicating the cooler temperatures inside each. On the front side of each kettle was a nozzle that appeared to fit the hoses attached to the cylinders. A door on one side of the room was labeled CAUTION BSL3 Facility. Saresh and his men probably had the capacity to produce the virus here too. The refrigerated truck was parked against the near wall next to the office that held Saresh and at least two other men. Jack looked around for something to start a fire. He didn't see any aviation fuel, but he did see a tank that was labeled 70% ethanol. The alcohol would definitely burn, if he could just find something to ignite it.

Sweat was starting to trickle down his back and his sides. He could feel drops starting to roll from his hair down his face. The sweat was partly from the intense heat inside the safety suit and partly from the nerves Jack felt were coming undone. His heart was beating so hard he could hear the thumping in his ears. He was right in the lion's den now, and he needed to slay the lion.

"Hey, you're back already?" Jack heard a muffled voice in front of him. He looked up and saw both men looking back at him. Quickly ducking his head back down so they wouldn't see his face, he grunted and nodded his head.

"Sucks. He's such a slave driver," the second man said. "Let's just get this one on the plane and then fuck him, we'll take a break. It's too damned hot."

Jack nodded his head and took a slow step forward. The two men were moving the cylinder toward the front of the barn in the direction of the plane. Where was Amy? He was beginning to question his trust in her. He had asked her to create a diversion when he got into the barn. If he got too close to these guys, they would surely notice he wasn't the man they thought he was.

Jack heard a woman's voice far behind him. He turned at the same time the other two men looked up. Standing on the far side of the fence was Amy wearing only her shorts and her button down shirt which she had fully unbuttoned. Every time the wind blew, the shirt opened up revealing her braless, very full breasts. Jack and the other two men were mesmerized by the sight, like men finding a watering hole mirage in the desert.

"Holy shit, Kevin! Let's go see what she wants."

The first man and Kevin started toward the fence while Jack feigned following them. Once they were in front of him he ripped his eyes from the beautiful sight at the fence and turned back into the barn. He looked around the interior of the barn but couldn't find anything to start a fire. The interior was quite spartan with a long, stainless steel waist-high table running the length of the room. The ethanol would burn, but Jack had nothing to ignite it. If only he were a smoker, he'd have a match or a lighter to use.

Next to the ethanol container was another large container with the label turned away from Jack. He saw that it was filled with phenol. Phenol was a caustic liquid that would kill any organism. It would have been used to clean off the suits and probably the inside of the virus canisters. At the bottom, Jack saw a nipple that would fit a hose for attaching the virus containers. If he could pump phenol into each of the canisters, he could destroy all the virus within. But he only had at most a few minutes to work.

Jack wheeled over one of the canisters and attached a hose first to the top of the virus canister and then to the bottom of the phenol tank. He pumped the handle a few times on the phenol tank to pressurize the inside, and then he opened the valve on the bottom of the tank. He could hear liquid rushing into the virus canister. He wasn't sure how much he needed to add to be sure of killing all the virus, but the didn't think it was very much. He waited only about 20 seconds before closing the valve and detaching the hose from the virus canister.

Jack pushed the first virus canister aside and pulled in the second one, repeating the procedure. Again, he pumped phenol into the canister for about twenty seconds before detaching the hose. He pushed both canisters back to their original position and grabbed for the third and final canister. He looked up and saw Kevin and his buddy running back to the barn. Jack knew he had to work fast to get this last canister filled with phenol.

The last canister hadn't been placed on a dolly yet, so he wouldn't be able to roll it over to the phenol tank. He tried to lift the virus canister up onto a dolly, but the tank was too heavy. He looked up and saw a block and tackle assembly. They must have used that to get the tanks onto the dolly for moving out to the plane. He didn't have time. He could see the two men were nearing the plane. Jack tipped the canister over on its side and rolled it over to the phenol tank. The loud rumbling noise would surely alert Saresh in the office, but he had no choice at this point. Sweat was pouring down his face, stinging his eyes, but he had to finish this one last tank.

Finally he had the tank over at the phenol tank. He reached out to pull the hose to the canister, but it wouldn't reach. The tank nozzle was too high and the hose too short to reach the canister. He would have to lift the canister to allow the hose to reach. He grabbed the tapered top end and pulled with all the muscles in his legs and back. The tanks slipped out of his gloved hands just as he was lifting it off the ground, causing a loud thud as it hit the ground.

"Hey, watch it!" Jack heard behind him in the direction of the office.

Jack ripped his gloves off to use his bare hands to grip the tank. He pulled again, finding strength he didn't even know he had. Slowly the tank began to rise off the floor as he felt every muscle fiber in his back strain at the weight. He got the tank into a standing position.

"Hey, where are Kevin and Josh?"

Jack didn't want to take time to answer this guy. He knew Kevin and Josh were on their way back into the barn and he had to hurry.

"What are you doing?"

Jack pushed the hose onto the canister and opened the valve on the phenol tank. He gave the handle a few quick pumps to make sure phenol would flow into the canister before turning around.

"Hey Saresh! Get out here!"

Kevin and Josh were screaming at the man in the door now and pointing to Jack. He wouldn't have time to clean up here; he had to get out now. Saresh was standing behind a man at the office door. Kevin and Josh were now running around to the other side of the plane. Jack looked around the barn again, but there was no other way out besides the large doors by the plane. He knew he had to act quickly. They would kill him for sure if they found out what he had done. They had nothing to lose; they were planning to kill a lot more people.

Chapter 40:

Jack sprinted across the barn and around the bench, heading toward the back of the plane away from the office. Kevin and Josh split up with one coming around the front of the plane and the other turning to head off Jack at the back of the plane. As soon as Jack had cleared the barn door, he turned to his right and away from the plane. He knew he had to outrun his pursuers and get over the fence in order to survive this. He had something to live for, someone who cared for him, someone who was waiting for him.

Jack got around the plane only a few feet before the one man came around the back. Running as fast as he could, he headed for the fence and out of the compound. He wouldn't be able to head toward Amy now, but he could try to find somewhere else to get over the fence. He knew this was a long shot, but he had to keep running. The full hood over his head was limiting his vision, allowing him to see only straight ahead. His breathing echoed inside the hood, sounding like a waterfall in his ears. He could hear the shouts behind him, but had no idea how far behind they were. He just knew he had to keep running. The shoe covers were making it difficult for him to keep his footing on the hard rock surfaces, so he tried to choose footfalls for the softer dirt even if this meant it would slow him up now. A fall would mean certain capture.

Jack could see the fence now although he didn't see Amy. Right then he heard a shot and saw the dirt explode to his right. People were shooting at him! He was a scientist. People shouldn't be beating him up or shooting at him. He shouldn't be running for his life. What was he doing? Another shot brought him back to reality. This one exploded to his left, and then he remembered his gun. He knew he wouldn't be able to shoot while running this hard. Hell, he probably couldn't even get the safety off while running this hard. The gun would probably slip right out of his hands he was sweating so much. The thump, thump in his ears from his pounding heart had become a load roar now, and he could feel the adrenaline pumping through his veins as he ran harder.

As Jack leaped over a small dry ravine, he could see the fence was only about 20 feet further. He still wasn't sure how he would climb over it. If only he had kept his gloves on, he could have used those to protect his hands from the barbed wire. He heard a shout from his right.

"Jack, Jack!"

Jack turned and saw Amy crouched beside the fence, pointing to a hole under the fence. He could see now that where the ravine crossed under the fence, there was a small gap that he could probably squeeze through. It was his only chance. He turned to his right and caught a quick glimpse back. He saw one man picking himself off the ground about 50 yards back and the other man running about 30 yards back. Jack jumped down into the ravine and raced for the hole in the fence. As he approached, he put his hands out and slid head first along the dirt hoping to slide under the fence.

He felt the skin on his hands rip as he hit the hard desert floor, and then he felt the pain in his stomach and under arms as he slid toward the fence. With the pain searing up from under him, he slid forward. He thought he was going to make it. He could see Amy crouching on the other side of the fence, with concern in her eyes as she looked past Jack at the onrushing men. Everything seemed to move much too slowly for Jack. He slid for what seemed an eternity with the fence moving closer ever so slowly, and then his head passed through the opening before he stopped. It felt as though he had been grabbed from behind, and he was sure he was about to be pulled back through the fence.

He grabbed at the rocks and dirt to scramble through the hole, and that's when he felt what had stopped him. The underside of the fence held his upper back in a death grip. The pain was unbearable as the short spiked ends of the chain link fence dug into the skin and muscles of his upper back like long nails from the fingers of Death himself. Jack pulled with his hands but only felt the fence dig in deeper.

"Hurry, hurry. They're almost here!"

"I'm, I'm stuck."

Jack reached around his back for his gun. He passed it through the opening to Amy.

"Here, shoot. It'll slow them down."

"I don't know how."

"Flip the safety off like I showed you. Just aim in their direction. It'll slow them down. Hurry!" Jack screamed as loud as he could knowing the hood was muffling his voice.

Jack pulled with his hands, grabbing onto dirt and weeds but nothing would give him the grasp he needed to pull himself out of the hole. He looked up and saw Amy squatting, with the gun pointed through the fence over his back. He saw her close her eyes and pull the trigger. The noise was deafening and the pain in Jack's ears made him think he may never hear again. He looked up again and saw Amy on her back, a victim of the unexpected recoil of the large caliber gun. She had a shocked look on her face as she got back up dropping the gun to the ground.

"I don't see them."

"They're probably ducking for cover. Pull me out, quick."

Amy reached down and grabbed both of Jack's hands, leaning against his weight, she pulled as hard as she could while Jack dug his feet into the soft dirt behind him to push. Jack pushed harder with his feet as he heard the protective suit rip against the exposed prongs of the fence. He was starting to slide under and continued to push with his feet as Amy pulled on his arms.

"Harder!"

They both heard the gunshot at the same instance the ground exploded next to Amy. Jack looked into her eyes and saw the terror mirroring his own feelings. Another gunshot ricocheted off a rock somewhere close by.

Jack dug his feet in for one final push and felt the fence rip its claws down the length of his back. Scrambling to clear the fence on the other side, he reached down for his gun.

"Get down!"

Jack took a stance with his legs positioned forward and back to provide stability for the recoil. Not taking the time to aim, Jack fired two shots in the direction from where he had just come.

"C'mon, let's go."

Jack got to his feet and half dragged and half pulled Amy to get her going. He tore off the hood of his protective suit and ran quickly at an angle away from the compound, but toward his motorcycle. He heard two more shots from behind him but didn't see how close they landed.

Jack stopped and turned, firing one more shot in the direction of his attackers before racing on behind Amy. It wasn't until they neared the Harley that they finally slowed their pace. Jack looked back but didn't see anyone following.

"They'll be coming after us." Jack looked at Amy and saw tears streaming down her face. Her hands were shaking uncontrollably. "It's okay, we're okay," Jack reassured her.

"But you didn't stop them. They still have the virus."

"I pumped phenol into three of the four canisters. The fourth was already on the plane. I detached the hose, but I didn't have much…"

Both heads turned as the sputtering sound of the airplane engine roared to life. They watched as the plane sped straight at them just clearing the fence before flying overhead. Jack looked up and saw one canister under each wing.

The site of those two bomb-like canisters on the wings of the plane made Jack think of Armageddon. If he hadn't destroyed all the virus, if he hadn't fully uncoupled the hose on the one canister he may be looking at the final deadly device to be used on the human race. He was looking at a sight a thousand times more deadly than the nuclear bomb that was dropped on Hiroshima. He was looking at the Enola Gay of the 21st century. And he had had a chance to stop it. That thought left his stomach churning.

The plane rose to an altitude of about 300 feet before banking to the south. The altitude was much higher than the average plane would fly for crop dusting, but Jack figured it was probably just about right to allow maximum dispersion to the west. As the plane leveled out on a southerly heading, Jack watched in horror as a spray of liquid exited and swirled behind the plane. He knew what that meant. Time seemed to stand still as the plane dumped its deadly load of virus into the wind. Jack and Amy stood rooted in one spot next to motorcycle, both with tears in their eyes.

Chapter 41:

A cloud of dust came swirling along the road from the west accompanied by a large rumbling sound. Jack watched as three Humvees drove down his small country road and pulled in at the gate of Intech's barn. The general had been mere minutes too late to stop the plane. Jack hopped on his Harley and felt Amy get on behind him. He rode back down the dirt path to the road and up to the gates feeling defeated. The capture of Saresh and his co-conspirators now would have little effect. The guards had been handcuffed and two Humvees were parked inside the compound near the barn. The general stepped out of the vehicle that was left at the road as Jack pulled up.

The general waved at Jack as he killed the engine and got off the bike. "The plane took off already General Osterman. It had two canisters, one under each wing."

"I know, I called in two Apache helicopters to escort the plane back here. Jack, I need to know, did the plane have the virus in it?"

"I can't be sure. I found three canisters and filled them with phenol to destroy the virus, but I can't be sure they didn't have other canisters."

"How did you find the canisters..."

"Jack." Jack turned back to the Humvee and saw Dr. O'Leary stepping out.

Jack turned back to the general. "I got into the barn and hooked the canisters up one at a time to a tank labeled phenol."

"That was either very courageous or very stupid, son."

"I know."

"Is there anyway you can determine if the plane released live virus?"

"What could you do if the plane did release the virus? We don't have a cure or treatment for it. It is as surely a death sentence as if the people of this city had just had a nuclear missile launched at them."

"We have contingency plans for situations like this. We can halt the spread of a biological weapon. I will quarantine the whole city. All air traffic and road traffic will be halted. We will send in the military to keep calm on the street. Martial law if you will."

"Well, if I could get a sample from the canisters, I could infect some mice or cells, but that would take a few days. Twenty-four hours at the least."

"I'll have to quarantine the city until those results are known then." The general called for one of his assistants.

"Wait General Osterman," Dr. O'Leary said. "We could get a sample and use the electron microscope at the institute. We could have an answer in a few hours."

"How will the electron microscope help you?"

Dr. O'Leary answered the general, "If the virus was neutralized with the phenol, we will be able to see the destroyed virus particles in the microscope."

The general appeared to think about that a few minutes. "I'll give you an hour after you land."

"Land?" Jack said.

"As soon as we get this plane in here, get a sample of the virus and then one of the Apaches will fly you to the institute. But you do realize that if this is infectious, you will not be coming back out of the quarantine zone?"

"I do, General, that is why I'll go alone," the professor said.

"John," Jack protested, "you don't know how to set up virus samples for the electron microscope. I'll come with you."

"No, Jack, you stay here. I helped create this thing. I'll go back into the city."

"I don't care. I'm going with you. With my help, we can get an answer much faster."

"Then I'm coming, too."

All three men looked at Amy who with her unbuttoned shirt showing off so much of her chest, looked little like the scientist she was. "Amy, there's no need for both of us to go."

"You know I can run that microscope better than either of you. I need to do this," Amy said as she looked pleadingly at Jack's.

Everyone turned their eyes to the sky to watch the Apache helicopters escorting the crop-dusting plane down onto the landing strip. Jack, Amy and the professor walked through the gates of the compound and around to the plane. Jack noticed that the canister he had unplugged had remained unplugged, so that was good news. The other wing was still dripping liquid at the nozzles. Jack took several small vials and swabbed a sample from each of the six nozzles. These were placed on ice and the three were hurried onto the Apache that had never let its rotors stop turning.

The ride back to the Institute took only fifteen minutes, but that was more than enough time for Jack to think about the disaster that the virus would have on this city. Worse would be the compounding effect of placing the whole city under military control and shutting down traffic into or out of the city. The panic would become enormous, especially once the first victims starting dying.

Brian J. Spencer

Jack could imagine the first victims might wake the next morning with a sniffle and maybe a cough. By midday they would probably experience a high fever from the increased replication of the virus, but would think it was only an early winter cold or flu. By the end of the day they would probably have difficulty breathing due to the fluid build up in the lungs from the action of the ricin toxin. If they made it to the hospital that night, they would only succeed in spreading the virus to more people before they died. There would be no treatment or cure for this disease.

Intech, with the help of scientists like himself, had developed the ultimate bioweapon; one that would bring citizens to their knees through death and panic. Jack wondered what Saresh must have thought when he realized the potential of the weapon he had in his hands. Did he really take it because he realized what the U.S. government was developing, or did he take it as a weapon that he could use for his own purposes? What really separated a scientist like Saresh from one like Jack? He hoped to one day have a chance to ask Saresh these questions.

Jack thought about Samantha. If he did have live virus with him now, he might never see her again. He should have gone back to tell her what was going on. He knew now he loved her, too. He had asked that the general see her and Brandon, but he knew he might never see her again himself. Still he realized he had made the only reasonable choice. This trip into the city might end his life, but it would provide the general with the information he needed to save Samantha's.

Chapter 42:

The helicopter came to a landing in the middle of the street that ran alongside the Hoffman Institute. The sight frightened motorists as well as the security guards at the institute. Jack, Amy and Dr. O'Leary hopped out and sprinted to the lab. Jack began his preparation of the virus sample for viewing under the electron microscope. The professor and Amy went to the microscope room to set up the machine and computer for viewing.

Preparing the sample was one of the most important steps before observing the virus. Although each virus family had a fairly unique shape, a scientist could almost never determine what kind of virus it was by using the electron microscope alone. Virus samples were prepared differently depending on the type of virus. All Jack was hoping to determine from this sample was whether or not the virus particles were still intact. If his plan was successful, the phenol should have denatured the protein shell of the virus and destroyed the normal shape of the virus.

The electron microscope at the institute was housed in a room all to itself. This device is extremely expensive; therefore, one microscope is shared among all those working in the institute. Regular light microscopes are not able to magnify an object as small as a virus. The wavelength of light limits the magnification to 1000 times making them useful for observing only bacteria and cell structures. However, the wavelength of electrons is much smaller and so higher levels of magnification can be obtained with this microscope, greater than 100,000 times. With this microscope, individual virus particles can be seen on the computer screen.

Jack checked his watch; it had taken him 45 minutes to prepare the virus samples. The general had given them one hour before he would order the quarantine. Jack ran down the hall to the microscope room and saw Amy sitting at the control console. Her hair was frizzled and standing out all over from the motorcycle ride. He could see that she had not had time to put her bra on and in fact had missed fastening several buttons at the top of her shirt. Jack had removed his protective suit, but the shredded remains of his shirt hung loosely on him. The wounds down his back still burned, but he pushed that into the back of his mind as he worked.

"Okay professor, here are the samples. Put the first slide on and Amy can get the images."

An agonizing five minutes passed before the first images came onto the screen. Jack never thought he would be looking at images that would determine his life and the lives of a million others. He had entered science so that he could help people using his research, but he never imagined that his decisions would affect the lives of so many people so directly.

All three were huddled around the computer screen as Amy typed furiously at the keyboard attempting to bring the image of the virus onto the screen. Jack felt his heart begin to race again as his mouth went dry. He steadied himself on the back of Amy's chair. Finally the screen went blank of all the toolbars and menus, and the microscope image came up. And what a beautiful image it was! Littered across the field of view was nothing but remnants of viruses. Virus capsids had been burst open and had spilled their contents out. They would no longer be capable of delivering their deadly payload.

Jack smiled as all three let out their breath in a long heavy sigh. Dr. O'Leary stood up and slapped Jack across his tender back sending waves of pain through his body again. "Good, let's get to the general."

The phone was already ringing as the three entered Dr. O'Leary's office where less than twenty-four hours earlier, Jack and Amy had been saved from certain death.

"Hello," the professor answered. "Yes, General, the virus was destroyed by Jack before the plane released it. That's right, we can do that. I will."

The professor handed the phone to Jack. "Yes sir?"

"Jack, I suppose what you did was heroic. You saved us all. I want to commend you for that."

"General, I have a request. I want to get back out to my cabin to spend a few days. I think I need a beer and then a long sleep. Could I get a ride back out in the chopper?"

Jack heard the general laugh, "A beer, Jack? I think you deserve a bottle of reserve Scotch, which I will hand, deliver to you when you get back to town. I think a ride back out here is the least I can do for you."

Jack couldn't wait to see Samantha again and recuperate at his cabin. He thought he would probably need more than a few days this time.

Chapter 43:

Jack glanced around the barn at Saresh who was handcuffed on a chair inside the office along with the other members of Intech. He wasn't saying anything yet, but the general assured everyone they would find out who had financed this operation and who else knew about this virus. Jack couldn't believe that something so deadly hadn't been developed by a terrorist nation. This virus had been developed by scientists that were funded by the United States government years after President Nixon had signed the treaty banning research on biological weapons.

Nothing had changed by the signing of that treaty. Instead of asking scientists to develop these new weapons of mass destruction, the government had recruited scientists by letting them believe they were working to better mankind by studying deadly diseases and toxins. How many scientists right now were working on a project they erroneously thought was to help humans conquer disease? How many of those people were secretly funded by DARPA for the benefit of the United States bio-warfare program?

Jack turned away from the congratulations and the conversations on what to do with a terrorist like Saresh. He walked slowly out of the barn and to his bike, which was still parked in the driveway. He hopped on jumping up to start the engine. As he sat listening to the tha-thump, tha-thump sound of the V-twin warming up, he was surprised to hear Amy's voice behind him.

"Jack, I ahh."

Jack turned back to see Amy standing on the road behind him. She had buttoned up her shirt and was making a feeble attempt at patting down her frizzled hair. She had stood up for him when it really counted that afternoon, putting her life on the line to get him out of the Intech compound. The time to heal those wounds was now.

"I know," Jack said. "I believe you. I'll see you next week at work."

Jack turned back to the road ahead and eased out the throttle with only one thing on his mind.

About the Author

Brian J. Spencer received his doctorate in medical microbiology and immunology from the University of Wisconsin- Madison. He is currently investigating the use of viral vectors for gene therapy at the Salk Institute for Biological Studies in La Jolla, California. Brian has published several papers in respected scientific journals on the subjects of gene therapy and anti-viral research. Brian lives with his wife, Jennifer, and two dogs, Ace and Chiquita, in San Diego.

www.ingramcontent.com/pod-product-compliance
Lightning Source LLC
Chambersburg PA
CBHW031825170526
45157CB00001B/185